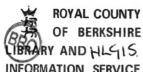

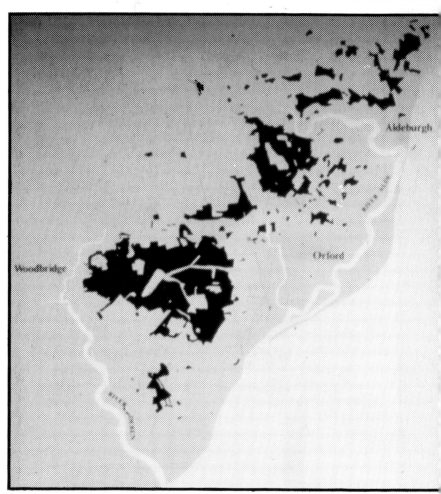

1881

THE LOSS OF HEATHLAND

SANDLINGS

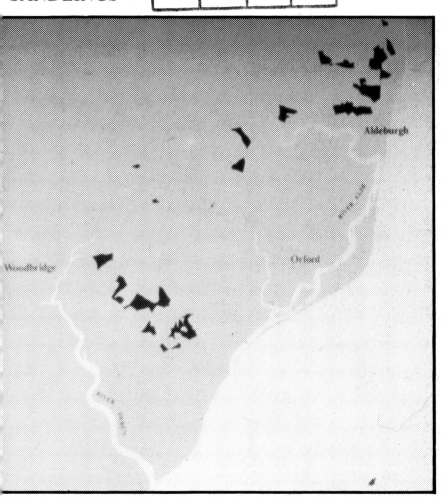

1981

OVER THE LAST HUNDRED YEARS

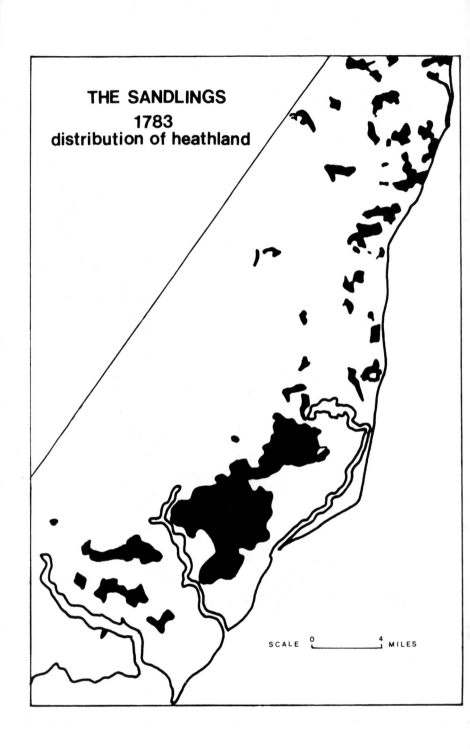

THE SANDLINGS
1783
distribution of heathland

SCALE 0 ⸺ 4 MILES

THE SANDLINGS 1931-32
distribution of heathland

based on land utilization survey map 1931-32

SCALE 0 — 4 MILES

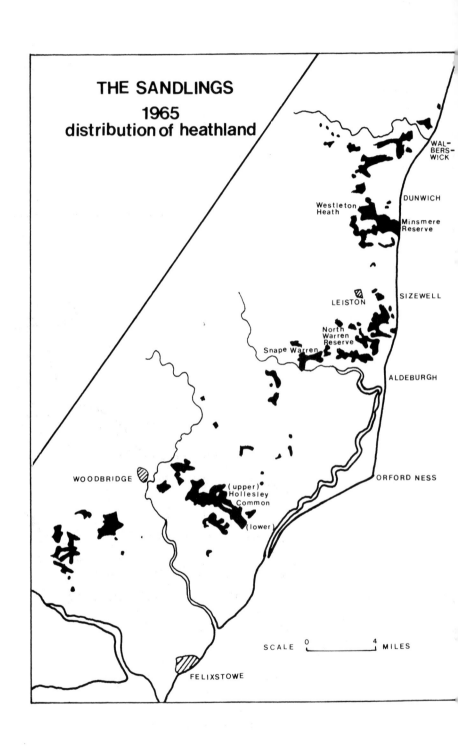

THE SANDLINGS

1965
distribution of heathland

WAL-
BERS-
WICK

DUNWICH

Westleton
Heath

Minsmere
Reserve

LEISTON

SIZEWELL

North
Warren
Reserve

Snape Warren

ALDEBURGH

WOODBRIDGE

(upper)
Hollesley
Common

ORFORD NESS

(lower)

SCALE   0   4   MILES

FELIXSTOWE

	a r p	a r p
t Hon.ble Lord Huntingfield *recives of* ann Tatnall Esq.r *following pieces of Land*		
57	. 3 . 27	
58	1 . . 22	
59	1 . 2 . 6	
50	6 . . 11	
51	13 . . 33	
60	3 . 1 . 22	26 . 1 . 7

N EXCHANGE

he following pieces of Land r Cottage given by him to ann Tatnall Esq.r

52	1 . 3 . 35	
55	. 3 . 7	
54	1 . 1 . 3	
55	. 3 . 20	
56	1 . 1 . 5	
49	. . 2 . 5	
61	. . 3 . 6	
8	12 . 3 . 4	
9	2 . 2 . 3	
59	8 . 3 . 30	
40	1 . 3 . 15	
41	1 . 3 . 15	51 . 1 . 33

56	7 . 2 . 11		Buck
57	13 . 1 . 16		
10	2 . 2 . 10		
52	4 . 3 . 7		
11	3 . . 27		
55	6 . 1 . 32		
12	2 . 2 . 22		
16	2 . 2 . 26		
17	1 . . 22		
58	. . 2 . 32		
15	1 . 2 . 44		
55	2 . 2 . 26		
21	3 . 1 . 32		
45	6 . 3 . 6		
14	1 . . 9		
54	2 . . 10		
18	4 . . .		
44	7 . 3 . 2	70 . 3 . 14	
	Total	271 . 2 . 11	

LEISTON COMMON 1824 ENCLOSURE AWARD

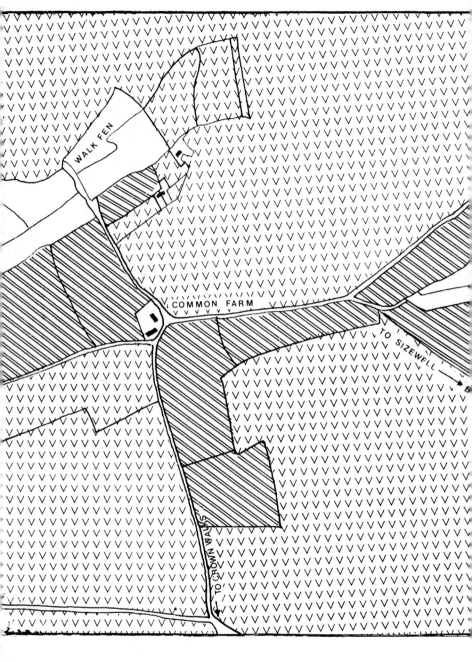

LEISTON COMMON AREA 1842
LAND USE BASED ON TITHE SURVEY AND MAP

SCALE 0   5   10   15 chains

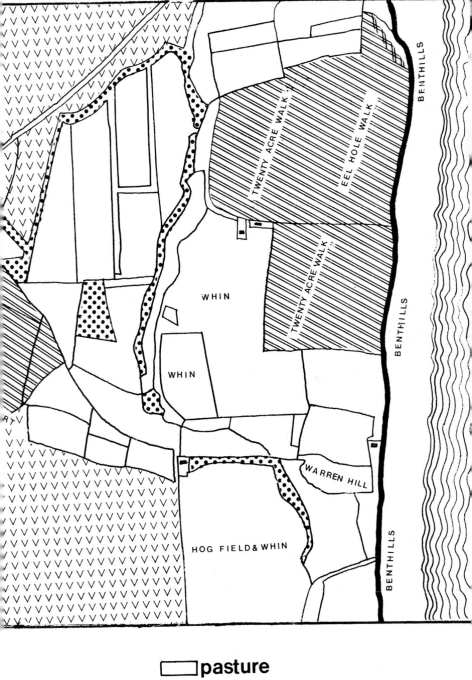

TWENTY ACRE WALK

EEL HOLE WALK

TWENTY ACRE WALK

BENTHILLS

BENTHILLS

BENTHILLS

BENTHILLS

WHIN

WHIN

WARREN HILL

HOG FIELD & WHIN

☐ pasture
▨ agriculture
ⅤⅤⅤ common ᴏʀ waste
▦ plantation

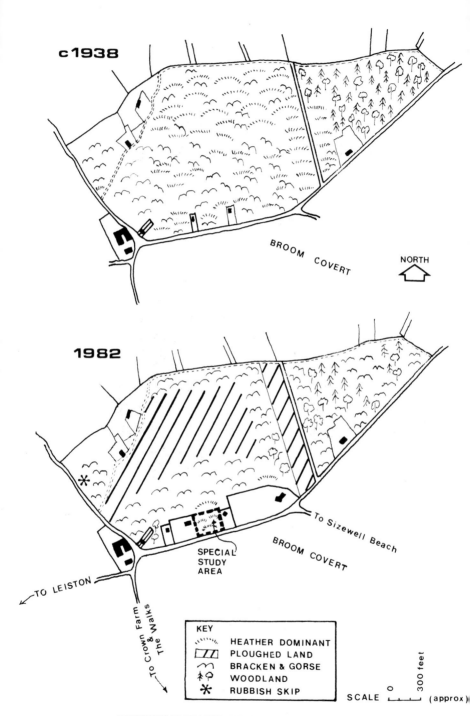

DECREASE IN HEATHLAND ON LEISTON COMMON 1938–1982

IN SEARCH OF HEATHLAND

Lee Chadwick

Dennis Dobson
London and Durham

For most reserves mentioned in the text, access permits
are required in view of the sensitive nature of the areas.
Application should be made to the Nature Conservancy
Council regional offices, a list of whose addresses is
available from Information and Library Services, Nature
Conservancy Council, Calthorpe House, Calthorpe Street,
Banbury, Oxfordshire OX16 3EX.

First published in Great Britain in 1982
by Dobson Books Ltd, Brancepeth Castle, Durham DH7 8DF

Photoset by Photobooks (Bristol) Ltd, Barton Manor,
St Philips, Bristol BS2 0RN

Colour folding plates printed by W. S. Cowell & Sons Ltd,
Buttermarket, Ipswich.

Printed and bound by Billings Ltd, Guildford, Worcester
and London

ISBN 0 234 72259 2 hardcover edition
ISBN 0 234 72260 6 paperback edition

FOREWORD

While moors owe their creation to mist and rain sweeping across hill country, heaths are the product of arid sands, wind-blown from beaches or spread long ago by glacier, torrent and wave. The plants and creatures of heathlands have learned to survive the special difficulties imposed by a harsh environment and have developed communities unique in character on this account, varying here and there as they reflect differences in local climate, aspect and land treatment over the centuries.

This book tells their story in fascinating breadth and detail, bringing under review a wealth of history, physical and human and describing the ways of life pursued by a diversity of plants, insects, reptiles, birds and other inhabitants of our remaining oases of heather, gorse and bracken. A long and patient study of East Anglia's still fairly rich heritage of heaths is the basis from which excursions for the appraisal of samples elsewhere in England are undertaken.

Something of the enchantment found by Jean Henri Fabre in his classic entomological researches conducted in the south of France a century ago is revived here for the delight of naturalists and it is clear that this is where the author's enthusiasm finds most heartfelt expression. At the same time, the ultimate fate of this unique feature of the English countryside must depend on establishing an enlightened concept of heathland as an amenity exploited and enjoyed by people of many persuasions through the ages and in our time. Lee Chadwick's book brings together all the threads of interest, as in a tapestry, so that we may view the whole scene in perspective in a way never attempted before with such success.

E. A. ELLIS

CONTENTS

NOTES ON THE ILLUSTRATORS

ED = EVANGELINE DICKSON

Landscape and natural history artist exhibiting regularly in London and Suffolk where she lives and works. Honorary warden for the Suffolk Trust for Nature Conservation on three nature reserves, she is particularly concerned by the increasing disappearance of our natural habitats.

BW = BARBARA WILLIAMS

Now living in Dorset, much of her life has been spent abroad where she is well known for her paintings of tropical birds and plants.

Painter in various media, she has illustrated books on several aspects of natural history.

PC = PAXTON CHADWICK

Designer and naturalist whose paintings for Malcolm Smith's *British Reptiles* in the King Penguin series first established his reputation as one of the foremost natural history illustrators. His death in 1961 left unfinished a major work, a commissioned flora of the British Isles. Most of his natural history work was done on the Sandlings of Suffolk, the study area which became the focus for this book.

DW = DOROTHY WOODS

PM = PETER MARTIN

ILLUSTRATIONS

15

ACKNOWLEDGEMENTS

In exploring the social and natural history of heathland, this book confines itself to the lowland heaths of southern Britain. It is to these lowland heaths that the term heathland generally applies, similar plant communities on the deeper peat of upland areas being usually known as moorland though the difference is not clear cut.

The principal areas of heathland are found on a number of cretaceous deposits of south eastern England, the Breckland and the east Suffolk Sandlings, on the Tertiary deposits of the Hampshire and London Basin, on the East Devon Commons, the Lizard and the Land's End peninsula. Rather than try to cover all these areas in brief and do justice to none, in the space available it seemed better to concentrate on home ground and then, travelling east to west, extend this personal search to selected representative areas of lowland heathland, even though this left unrecorded visits to such fine places as the heaths of Ashdown Forest and the excellently documented Hothfield Common Nature Reserve in Kent.

What must be recorded is the most kind co-operation received in all areas from the Nature Conservancy Council regional offices, County Conservation Trusts and Naturalist Societies, The National Trust and in particular to the Reserve wardens mentioned in the text.

The author wishes to acknowledge her debt to all those very busy people who nevertheless gave so generously of their specialist knowledge and advice which alone made this venture possible.

For advising on Part I, special thanks are due to Dr Joan Thirsk, Mr E. P. Thompson, Prof. Winifred Pennington Tutin, Dr C. J. Wainwright, and Mr G. Ewart Evans. Also to Miss E. D. R. Burrell for kindly allowing the use of material in her unpublished M.Sc. thesis *The Sandlings of Suffolk 1600–1850*, to the Suffolk County Planning Officer for information supplied by his department and to Mr G. S. Ogilvie for details concerning the historical background of F. M. Ogilvie's *Field Observations of British Birds*.

For advising on Parts II and III including identification of species, the author is specially grateful to the Institute of Terrestrial Ecology for help afforded by the staff of Furzebrook Research Station, and to the Trustees of the British Museum (Natural History) for the help given by the staff of the Entomology and Botanical Departments as well as to Dr E. A. Ellis, Mr H. E. Chipperfield, Mr S. Beaufoy, Dr Lewis C. Frost, Mr E. Milne-Redhead, Mr P. J. O. Trist, Mr P. Lawson, Mrs S. M. Turk, Mr R. D. Penhallurick, Mr W. H. Payn, Mr

A. E. Axell, Mr J. Sorenson, Mr J. Grant. For identifying specimens and for other assistance, grateful thanks are also due to the Royal Scottish Museum (Natural History Department), the Castle Museum, Norwich, and the Ipswich Museum (Departments of Natural History and Geology), and the Mycological Section of the Herbarium of the Royal Botanic Gardens at Edinburgh and Kew.

Finally, the author wishes to express her appreciation to the artists who in illustrating this book from "the living heath" have given unstintingly of time and energy to produce illustrations whose quality speaks for itself. Of special value was the collaboration with Evangeline Dickson in visits made to the west country, enabling author and artist to explore together the heathlands and so achieve the essential unity of text and illustration.

NOTE ON SIZES

In some species there is considerable range in size and often a difference in size between sexes. Unless otherwise stated, measurements are from specimens drawn.

Abbreviations
BL=Body length
WS=Wing span
Enl.=Enlarged.

Introduction

OUR VANISHING HEATHLANDS

BLACKBERRYING

"That particularly British institution of the Common." The phrase occurs in an unsigned article in *Leisure Hours*, a Victorian family magazine for the year 1891, in which the author exclaims:

> Waste land!—a term not infrequently applied to these stretches of country which know not the bright blade of the plough; waste land forsooth, where man can breathe the pure air of heaven free from contamination of his own greed, where he can roam at will, getting from nature in spite of himself, restful ease with soothing influences. Surely here is not only waste. Here in England that men should have fought, are fighting still and will no doubt continue to fight for these few acres of waste land, is an index to the character which gave origin to the Great Charter . . . There is scarce a stretch of common land in England but what has been a battle field; a bloodless one tis true, a struggle in many instances of the weak against the powerful. Now the loiterer wanders unaware of the unsung warriors who fought for the blessed privilege of the right to a few acres of the earth's surface . . . What a glorious sense of freedom is here, what fellowship with bird and beast, what a companionship even in the coarse, wild growths!

The sentiment, if not the style, of this Victorian enthusiast will find approval today among those who share his passion for open heathland. In the intervening ninety years or so, thousands of acres of heathland have disappeared. Some have gone under the plough for now new methods of agriculture enable heathland to be reclaimed profitably. Some have been transformed into dark forests of home grown timber. Some have been converted into housing estates to meet the needs of population dispersal. In 1960 commons of all kinds occupied some 4.03% of the total land surface of England and Wales, some 1,500,000 acres (607,000 hectares) which was little less than the total area used for houses and gardens at the same date.

During the second world war, 21,000 acres (8,500 hectares) of commonland in Britain were requisitioned for food production and some produced excellent crops. However, even in recent years extensive acres of heathland, ploughed up with insufficient knowledge of the land, sometimes in order to obtain a subsidy, have had to be abandoned as they proved unprofitable. Nor is this surprising for heathland is a term to describe a plant community connected with some of our poorest soils.

Within the apparent uniformity of heathland is a fascinating variety for the associated plants and animals vary according to the geography of the heath, the climate and soil. They are also very much affected by the treatment their home or habitat receives from people and animals. Some of our high upland areas are almost untouched by human kind, where a community of typical heath plants represents what is known as a natural climax vegetation. Left to itself, this is what nature produces because harsh weather and soil conditions discourage the growth of trees. Our lowland heaths, however, where conditions are suitable for tree growth are a different matter. Many probably owe their existence to the activity of human beings and their grazing animals, ever since primitive man settled in the less heavily wooded areas and kept the forest at bay. In fact, today where these activities have ceased altogether, heaths in places have changed to impenetrable jungles, cleared from time to time by fierce unplanned fires and it is thought that here lies one cause of some disturbing changes in the abundance of certain typical heathland plants and animals.

At present one might say that the whole future of our heathlands is in the melting pot and there is a danger that unless citizens are sufficiently stimulated to sit up and take notice, the "particularly British institution of the common" may be in greater danger, despite the recent spate of legislation, than at any time since the enclosure acts.

As is well known, from time immemorial some common lands have escaped every kind of pressure because of certain ancient folk rights which allow a commoner to take some portion of the natural produce of the soil belonging to another. Among these rights are the common of turbary or the right to dig turf or peat for use as a fuel in the home; the common of estover or the right to cut furze and bracken for firing or bedding for animals; the common of piscary or the right to fish; pannage or the right to allow pigs to rummage for acorns, beech nuts and fungi; "common in the soil" or the right to take gravel and perhaps the most important and widespread of all common rights— the right to pasture beasts on heath, marsh and meadow.

Before the Royal Commission on Commonland presented its report in 1958, it interviewed many people who claimed one or another of these rights. Following upon this, the Commons Registration Act of 1965 called upon local people to register commons, village greens etc. in their areas on which such claims could be made. Under this Act, each county maintains a register of commons that is open to inspection.

FURZE CUTTER

Seven years were allowed for objections to proposals to register land, but the rate at which disputes are being considered means that of 26,628 objections received in 1972 there were, in May 1978, 26,400 disputed registrations together with 4,800 claims to ownership still unsettled. In June 1981 a backlog of at least 15,000 objections concerning rights of way are still unheard and likely to remain so if the Wildlife and Countryside Bill before the House of Commons becomes law.

The Commons, Open Spaces and Footpath Preservation Society has pointed out that registration was only one of the three principal recommendations of the Royal Commission. The other two urged firstly that a legal right of access should be granted to all commons and secondly that wider powers of management should be given to local authorities and others to ensure that this right is not abused.

These provisions are immensely important because even when commonland is registered, this does not automatically give the public right of free access. All land in Britain has an owner and commonland is no exception although sometimes no one knows who the owner is. A growing number of commons belong to public bodies such as local authorities, Nature Conservancy Council, National Trust, the Crown Estate Commissioners and others, but many rural commons in particular are privately owned often by the local squire. Chiefly thanks to the Commons Preservation Society during the hundred years of its existence, a public right of access for "air and exercise" has been granted under statutory powers to some 400,000 acres (162,000 hectares) of commonland mostly in built up areas, but still a great deal of commonland—over a million acres—is not open to the public of right, but on sufferance only.

How correct the unknown writer was to query the term "wasteland". Some of these heaths and commons are of priceless scientific value because they provide a semi-natural habitat with a character all its own. The Nature Conservancy Council has urged that areas of these "wastelands" should be conserved as national nature reserves, so that our children's children have somewhere to carry out research into their unique community of plants and animals, some of which would cease to exist if their particular environment disappeared.

Even a human being living on a heath can grow to depend on all it has to offer. One becomes attuned to the space and quality of light, to the massed colour of the heather, gorse and bracken, even to the hot sandy soil underfoot. This book attempts to tell how one person who has become addicted to life on a Suffolk coastal heath set out to explore some different types of British lowland heaths and, at a time when scientists of many nations are looking closely at this interesting community of plants and animals, to glean something of the factors affecting its survival.

Part I

PEOPLE AND HEATHLAND

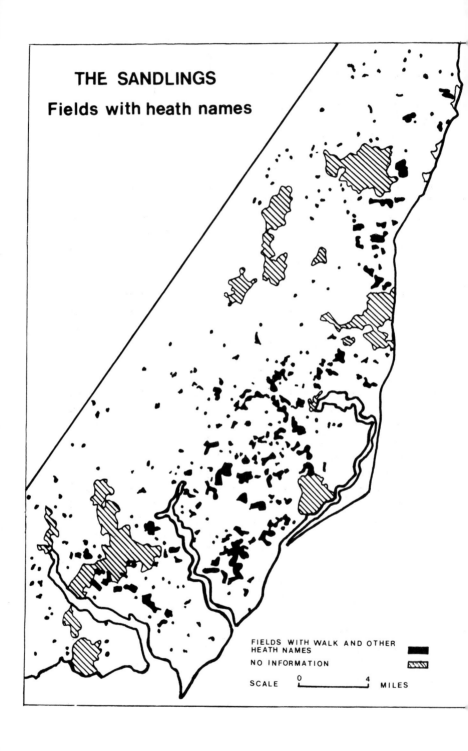

THE SANDLINGS

Fields with heath names

FIELDS WITH WALK AND OTHER
HEATH NAMES

NO INFORMATION

SCALE 0   4 MILES

A COMMONS SURVEY

When the Royal Commission on Commonland made the recommendation that wider powers should be given to local authorities and others, requiring them to set up statutory management committees to look after commons and draw up approved management schemes, it must have had in mind that for these plans to be successful they needed to be based on a wide range of relevant knowledge and, where possible, some sort of preliminary survey.

A start was made in this direction in 1960–65 when a series of field studies was carried out on the condition of 526 representative commons from every county except one in England and Wales. (Scotland has no commons.) Under the auspices of the Nuffield Foundation, the initiative came mainly from E. M. Nicholson, then Director General of the Nature Conservancy, with the support of a research team from the Department of Land Economy at the University of Cambridge and a widely representative advisory committee. Denman, Roberts and Simms (1967) summarised their conclusions in *Commons and Village Greens* with the aim of "helping anyone who set about the job of managing a common".

On the aspect of commons as places for refreshment and enjoyment, the team shed some interesting sidelights including the fact, for instance, that since bracken fronds reach their maximum growth during peak holiday periods, ever increasing tracts of bracken lacking effective management mean that large areas are put out of the reach of the public that otherwise might provide fresh air and exercise with least harm to the habitat.

Management problems were severe in some of the forty-two scrub commons visited where in the absence of grazing, gorse and bramble had taken over but here again the team mentioned as an important consideration the high value sections of the public attach to such unfenced open spaces that approach a wilderness or "state of nature" in which they are free to wander. Litter nuisance appeared to be less acute on rural than on town commons even where the density of visitors was parallel. In all commons visited, except in one major case, it was found that the use of a common for natural history and research purposes had not seriously clashed with its use for general recreational purposes.

For the purpose of the survey, commons were classified with an eye to managerial requirements as upland grass moors, upland heather and grouse moors, bracken tracts, lowland heather tracts, downland

and grass heaths, scrub commons, urban and metropolitan commons. There was also a little group of commons with such specialised purposes as fuel and poor allotments, gravel pits, quarries and mineral workings, turbaries, roadside verges and village greens.

The list reveals the wide variety of habitat encompassed in the term *common*. However, the popular conception of a common probably approximates most closely to the sentiments expressed throughout the ages in an assortment of nostalgic verse in which heather is an essential ingredient, typified by these lines from *In City Streets* by Ada Smith (*c.* 1860):

> Yonder in the heather, there's a bed for sleeping,
> Drink for one athirst, ripe blackberries to eat,
> Yonder in the sun, the merry hares go leaping,
> And the pool is clear for travel wearied feet.

Probably from very early times, heather has had a place not only in the affections but in the work-a-day life of people where its various parts have been put to good use: the strong wiry twigs for brushes and mattress filling, or piled high, sod upon sod, for walls, the flowers for making heather ale or tea infusions against the gout, and the juices for tanning leather.

46% of the commons surveyed were lowland heather tracts. Such lowland heather commons are widely distributed in thirteen English and seven Welsh counties and constitute a habitat whose conservation is considered of great ecological importance. In this connection it cannot be too often stressed that conservation is not only, or mainly concerned with rare species of interest to specialists. Of great importance is the conservation of natural and semi-natural communities of plants and animals known to us all, including those quite commonplace plants like the dwarf shrubs of the heather family whose presence in large numbers is characteristic of certain conditions that form a suitable place to live—an ecological niche—for a host of associated plants and animals.

HEATH DWELLERS IN LITERATURE AND LIFE

The word *ecology* is derived from a Greek word *oikos* meaning a home and those of us who share heathland as a home with all its other denizens in a real sense form part of the heathland community, having both a short term and a long term effect on it in a myriad of ways. Without perhaps being aware in detail of all the interaction between soil and climate, plant and animal life, human inhabitants too are strongly affected by the heath's total presence and this is reflected in the special place heathland holds in literature and folk-song. As Thomas Hardy (1895) says of Clym Yeobright whose home from childhood was a typical lowland Dorset heath:

> He was permeated with its scenes, with its substance, with its odours . . . He might be said to be its product . . . his estimate of life had been coloured by it.

Born almost a century before Thomas Hardy, in the England of George II, the artist craftsman Thomas Bewick was well aware that he was a product of his northern moorland upbringing. Much heathland was converted to arable in Bewick's lifetime as part of the speed-up of enclosures from 1760 onwards and he writes with longing of *Cherryburn* his childhood home where he was born in 1753 in the ancient parish of Ovingham in the Tyne Valley:

> To the westward, adjoining the house, lay a common or fell, which extended some few miles in lengths, and was of various breadths. It was mostly fine greensward or pasturage, broken or divided, indeed, with clumps of blossomed whins, foxgloves, fern and some junipers, and with heather in profusion, sufficient to scent the whole air . . . On the common,—the poor man's heritage for ages past, where he kept a few sheep, or a kyloe cow, perhaps a flock of geese but mostly a stock of beehives—it was with infinite pleasure I long beheld this beautiful wild scenery which was there exhibited, and it was with the opposite feeling of regret that I now find it all swept away.

In his *Memoirs* (1862) Bewick recalls how during his boyhood here and there on the heath adjoining his home could be seen "the cottage or rather the hovel of some labouring man, built at his own expense and mainly with his own hands." Of these fell-side neighbours he writes,

> These inhabitants of the fells and wastes whose cottages were surrounded with whin and heather I must observe that they appeared to me not withstanding their apparent poverty, to enjoy health and happiness to a degree surpassing that of most other men.

In contrast to this view, the reason given for the harsh Settlement Laws of 1662 was that

> Poor people are not restrained from going from one parish to another, and therefore do endeavour to settle themselves in those parishes where there is best stock, the biggest commons, and wastes to build cottages and the most wood for them to burn and destroy; and when they have consumed it, then to another parish; and at last become rogues and vagabonds.

Bewick however was obviously intrigued by the originality and independence of some of these heath dwellers, descendants of the cotters and bordars of old. There was Tom Foster, bee keeper, who hid his bee hives in the shelter of an old whin rush and made a good living from the common, but most typical perhaps of the squatters on the heath was Anthony Liddle who treated all men as equals—a singular personality whose whole cast of character was formed by the Bible that he had read from cover to cover.

> Acts of Parliament which appeared to clash with the laws laid down in it, as the Word of God, he treated with contempt and since he maintained that the fowls of the air and the fish of the sea were free for all men, consequently the game laws had no weight with him.

This feeling that the resources of nature should not be the individual property of any person, lord or not, is an attitude to nature which frequently appears in peasant movements of all sorts. (See

Rodney Hilton's *Bond Men Made Free* 1977.) William Morris in *The Dream of John Ball* speaks of "The struggle against tyranny for the freedom of life; how that the wildwood and the heath, despite wind and weather were better for the free man than the court or the cheaping town . . ." Bewick's fell-side neighbour was part of his long tradition going back five hundred years and more to John Ball, leader of the great peasant revolt of 1381, with his vision that all men in the beginning were created equal and "Matters goeth not well to pass in England nor shall do till everything be common and that there be no villein nor gentleman but that all may be equal."

Be that as it may, by the end of the eighteenth century there were many among Bewick's contemporaries who argued that the enclosing of waste lands was necessary even if it did not lead to extra profit because "the men who usually reside near a common, are the depredators of the neighbourhood; smugglers, sheep dealers, horse-jockies and jobbers of every denomination here find their abode." (Footnote by J. H. to Young's *General View of the Agriculture of the County of Suffolk* 1794)

TRAVELLERS IN THE SANDLANDS

Our own close association with heath and common began in the nineteen thirties when we made our studio home on the fragmented piece of coastal heath known as the Leiston cum Sizewell commons, part of the Sandlings or Sandlands of Suffolk, a name used in travelogues of the eighteenth century for the glacial sands and gravels giving rise to the heaths and marshes that now spread in discontinuous tracts along the east coast from Aldeburgh to Great Yarmouth. According to W. G. Hoskins and L. Dudley Stamp (1963) half to two-thirds of what was then East Suffolk's open heathland lies in this area though by no means all of it is common, even when so called. (Leiston though not Sizewell Common falls into this category of a common now only in name.) Of eighty-two commons listed totalling 4222 acres (1709 hectares) all but six are placed by the authors in the Sandlings belt.

In *The Suffolk Traveller*, which he published soon after making a series of surveys in the county, John Kirby (1735) wrote of Suffolk, home of the Southern Folk:

It is a maritime county bounded on the East by the Ocean . . . The county may naturally be considered as consisting of three different sorts of land viz. The Sandlands, the Woodland and the

Fielding. The Sandland is the tract of land which reaches the River Orwell, by the sea coast to Yarmouth and is pretty nearly separated from the Woodlands by the great Road leading from Ipswich through Saxmundham and Beccles to Yarmouth. This part may be divided into Marsh, Arable and Heathland.

Rather later in the century, a famous French traveller, François de la Rochefoucauld (1784) described his journey in the Sandlings to the region of Sutton Heath where "The further we went the deeper we found ourselves in soft sand." He noticed in this coastal stretch, in contrast to central Suffolk, the land was little enclosed except immediately round villages. He noted, too, that carrots were a feature of the agriculture of this corner of Suffolk and remarked on the fine breed of Suffolk horses found in most perfection in the district of the county upon the coast, "the best of all being found upon the Sandlings, south of Woodbridge and Orford."

The Frenchman was accompanied on his journey by a Suffolk farmer, Arthur Young, who later became secretary to the Board of Agriculture on its formation in 1793. In the account of his own travels *A farmer's tour through the East of England* 1771, Young greatly admires the farming husbandry in the triangle of land formed by the three points of Woodbridge, Orford and Bawdsey, but of the maritime region, now designated Heritage Coast, where occur the heath lands of which Leiston and Sizewell commons form part, he wrote: "There is a large amount of poor and even blowing sand to be found . . . The whole of the maritime region may be termed sandy." Suffolk's climate he describes as "unquestionably one of the dryest climates in the kingdom. The frosts are severe and the N. E. winds sharp and prevalent." At the beginning of the previous century, R. Reyce (1618) in his *Breviary of Suffolk* had described its fairer face: "Though it is not in all places alike, yett it is commonly esteemed that the air is as sweet and healthfull generally, as in any other county whatsoever."

Yet another eighteenth-century traveller, Daniel Defoe in *A Tour through England and Wales 1724*, refers to smuggling as amounting to a major industry in the south of England and certainly the coastal heaths of south east Suffolk were no exception. Four miles north along the coast from Aldeburgh, the small hamlet of Sizewell Gap was a notorious centre for the illicit trade, the extent of which at the end of the eighteenth century is described in the diary of surgeon-farmer William Goodwin of Street Farm, Earl Soham. Entries commenced in the freezing February of 1785 when he saw passing through his village from Sizewell Bay 2,500 gallons of smuggled spirits in twenty carts in six days. "The number of Horses on the Beach at Sizewell was frequently from 100 to 300 and of wagons and carts from 40–100 at a time." he recorded. One can well understand how important was a backcloth of heath and fen to "star-light trading" on such a scale.

Then as now, heathland and marsh ran down to the dunes fringing what was then a sandy beach of the North Sea at Sizewell Gap. Today the north-south drift of the tide and the scour of the waves have ground away the sandy rock that once stretched almost a mile further out to sea, so that today there are no high cliffs to break the level lines of sky and sea and dune but in 1874 White was noting in his *Gazetteer and Directory of Suffolk* "At Sizewell Gap, there is a fishing boat and coastguard station and the cliffs rise precipitously from the sea." The gap in these cliffs formed the scenario for smuggling operations covering a wide area in which "a desolate common of wide extent" acted as an escape route. Vaults in which to store the contraband goods were dug in the sandy soil, their contents covered with stout planks and the turfs replaced so that all was camouflaged by heath and bracken.

HEATHS AS SHEEPWALK

The complex of commons stretching along the coast from Aldringham-cum-Thorpe through Sizewell to Westleton and Dunwich lies in the Blything Hundred and White in his *Directory* (1874) notes that "the eastern side in the Blything Hundred extending some 60 miles along the coast . . . still has some unenclosed sheep walks." Many fields today, as the map of the Sandlings on page 26 shows, have names containing the term *walk*, like the Shepherd's Walk and the Lord's Walk on Leiston Common, indicating they were once open heath where sheep grazed.

Of the many-sided relationship between people and heathland, historically the most intimate has been that between the shepherd and sheepwalk—a connection reflected in some of the local names of the small flowers that grow in the cropped turf: shepherd's bedstraw *Asperula cynanchica*, shepherd's calendar *Anagallis arvensis*, shepherd's cress *Teesdalia nudicaulis*, shepherd's knot *Potentilla erecta* and sheep's bit *Jasione montana*. As the Commentator puts it in *The Shepherd of Bunbury's Rules* by John Claridge, Shepherd, which first appeared as *The Shepherd's Legacy* in 1670:

> The Shepherd, whose business it is to observe what has reference to the Flock under his Care, who spends his Days and Many of his Nights in the open Air, and under the wide-spread canopy of Heaven, . . . Everything in Time becomes to him a sort of weather gage. The Sun, the Moon, the Stars, the Clouds, the Winds, the Mists, the Trees, the Flowers, the Herbs and almost every Animal with which he is accounted. All these I say become in such persons Instruments of real Knowledge.

It is this "real knowledge" that one is conscious of when talking to Arthur Sutton, born 1904 in a cottage on the heathlands of Pickenham, Norfolk and now living in the village of Orford at the mouth of the River Ore. He was one of the last shepherds in this area of the Suffolk Sandlings when he retired from work ten years ago. The shepherd's hut he used, a handsome and strongly constructed wooden structure on seven inch broad iron wheels made to stand up to the hurricane force winds of the east coast, still stands as illustrated by the artist on the estate of the manor whose flock he tended. Each year around lambing time in early March, it was towed out to Snape and Tunstall heaths where it would remain for six weeks until lambing was over. (These are the heaths that before the forestry

plantations of the 1920's were referred to by the brothers Rainbird (1849) as "the large extent of barren soils and black heaths in the neighbourhood of Orford and Woodbridge.")

According to Arthur Sutton, the site chosen for the shepherd's hut was a rise on the heath, and if possible against a wood for shelter from the north wind. "Sheep always go to high land—a good instinct for if snow set in and you get a drift with them in a hole, then you have had it." Two men took it in turn to be out on the heath, each with a wooden sleeping bunk built into the wall of the hut. A straw yard was made for the dogs and a shelter for the ewes. If a lamb was sickly it was put inside a sack filled with soft hay, then brought inside near the fire to be doctored if necessary from the shepherd's store of remedies.

During the first world war, Arthur at twelve years of age accompanied his father with the sheep, earning as shepherd's page 2/6d a week day and 3/6d on Sunday. Most of the year at 7 a.m. he walked the sheep out over the heath to graze then brought them in about 4.30 p.m. to fold on turnips, kale or grass.

Sutton's father was a shepherd in the days when the tough buttonless working smock of the traditional kind was still worn and hand shears were used for sheep clipping. Shearing started about the first week in June just as the grease in the wool began to rise with the warmth. A team of clippers, ten at a time, used to go round from farm to farm where a welcome of home brewed beer usually awaited them. These were highly skilled men who even when merry had no need of Thomas Tusser's warning rhyme written for Suffolk farmers over four centuries ago as part of his *June Husbandrie*:

> Reward not thy shepe when you take off his cote,
> With two or three raw patches as broad as a grote,
> The flie then and wormes will compel it to pine
> More paine to thy cattell, more trouble is thine.

The two commonest sheep in Suffolk, sometimes referred to as heath sheep, were the Southdown and the Norfolk, both black faced breeds. On the Sandlings these were later crossed and re-crossed to produce the famous Suffolk Black-face with the hardiness and the fine flavour of the Norfolk and the early maturing qualities of the Southdown.

Of the quality of Suffolk sheep, R. Reyce was writing enthusiastically in his *Breviary* at the beginning of the seventeenth century: "Fleece is fine, flesh is sweet, so fatt and ready to breed and ever bringing profit to the owner as is well known in those parts next the

Champaigne." (Champaigne meaning unenclosed country.) In the seventeenth century, when cattle rearing and dairying extended from the boulder clay region of High Suffolk along the estuaries into the coastal area, "great beasts" were evidently allowed on to the heaths as well as sheep. The commons of Dunwich, for instance, were so managed that townsmen paid a yearly sum of money for the pasturage of both cattle and sheep on the heath and the marshes. The number that each was allowed to graze at any one time was graded according to their standing in the town. Bailiffs were allowed 3 cattle and 24 sheep, burgesses 2 cattle and 16 sheep and the remaining freemen only 1 great beast and 8 sheep. (Document 134.18.Jas.1 (1620), Suffolk Record Office, quoted by Burrell 1960.)

Surveys and maps for the apportionment of tithes between 1600 and 1842 show that much land in this region of heath and sheepwalk of the Sandlings was owned by rich yeomen who were a feature of south-east Suffolk—as well as by such large landowners as Lord Rendlesham and Sir John Blois. On these manorial estates, the lord of the manor usually kept some sheepwalk for his own needs as part of his demesne and rented the rest to his tenant farmers. In addition there were comparatively small areas of commonland on which the remaining villagers had grazing and other rights.

The importance of sheep and therefore of sheepwalk and heathy commons to the husbandry of the Sandlings of Suffolk in the eighteenth century is shown very clearly in the terms proposed by Sir John Blois to his tenant for the lease of one of three farms on his Walberswick and Blythburgh estates. According to an agent's description presented for valuation in 1769, there were about 1500 acres of unenclosed sheepwalk. In a document deposited in the Suffolk Record Office among the Blois family archives dated July 7th 1770, the lord of the manor stipulates that:

The tenant shall keep at least 800 sheep and shall fold same regularly at all reasonable Times in the Year, with the usual number of hurdles upon some part of the premises that shall likely to be benefited therebye, under the penalty of £5 a Night for every Night the Flock or at least 600 thereof shall not be folded upon some proper part of the premises Unless the same shall be kept out of the fold by Reason of badness of the Weather, and about two months in Time of Lambing.

It was this farm when Arthur Young visited it in 1779 that was growing nearly 300 acres of turnips annually as a fodder crop for "the black faced gentry".

Despite an increasing trend towards arable farming in the nineteenth century, John Glyde's list of occupations in *The New Suffolk Garland* 1866 shows shepherds in Suffolk increased from 583 in 1851 to 878 in 1861. The beneficial effects of sheep grazing on the management of heathland is well described in the following from White's *History and Gazetteer* at the end of the last century:

In the light sandy soils of Suffolk, breeding and close folding on the arable lands at night (of the flock) which has browzed on the heathland during the day is a system chiefly relied upon for fertilising the arable land . . . The heath or sheepwalk of which much of the district is covered is sandy gravel with no apparent depth of soil above it . . . covered with velvety herbage or moss or studdied more or less with broom and whin bushes. In other places it is entirely covered with 'ling' or heather . . . The only way to keep these large tracts of sterile sand to account is by keeping breeding ewes. In the day time, soon as the lambs are weaned, the ewes are made to browze on the long shoots of the whin bushes and pick the scant herbage of velvety turf that grows between. At night they are folded on the arable lands, bare fallow, old layers, or turnips as the case may be.

The velvety tread of "the sheepwalk's slender grass" was one reason why the Aldeburgh poet George Crabbe "Loved in the Summer on the Heath to Walk." Crabbe probably knew every inch of the "heath and common wide" that stretches inland along the coast between Aldeburgh and Dunwich. In The Borough (1810) he expresses the opinion that one of the sweetest moments life has to yield belongs to those who are fortunate enough to

Stray over the Heath in all its purple bloom—
And pick the Blossom where the wild-bees hum;
And through the broomy Bound with ease they pass
And press the sandy Sheep-walk's slender grass
Where dwarfish flowers among the gorse are spread
And the Lamb browzes by the Linnet's bed.

Nearly two hundred years later, dwarfish flowers, gorse, broom and linnet still abound even on this relic heath. The sheep are gone. Bracken and bramble have encroached on heather but myxomatosis has nevertheless left enough rabbits to keep down the coarser grasses near the paths where the turf still remains springy to the tread. Pheasant orientated gamekeepers may seek to contain the wanderer but for blackberry pickers and ramblers alike, straying on the heath remains one of life's good moments. And when at night under the huge globe of sky, the dark clumps of gorse and weird shapes of tangled briar crouch like sleeping beasts as the moon climbs out of the sea, the stillness seems to stretch back to the very beginnings of time, a sensation strangely enhanced by the distant hum of Sizewell's nuclear power station, floating eerily luminous over the benthills at the sea's edge.

HEATHS REVEAL THEIR HISTORY

The heathlands of East Anglia have provided clues to their own origins as well as to some very early patterns of human society. Scattered in their light soils, one comes across ancient flint instruments, among them small choppers and hand axes which the stone age farmer many thousands of years ago probably used to make clearances in the birch, pine and oak of the ancient woodland, thus beginning to create those open conditions it is now thought helped to encourage the formation of heathland.

The palaeolithic or old stone age people who fashioned the oldest flint artifacts were hunters and fruit gatherers, probably having little effect on the environment, and pollen deposits at this time point to a fairly uniform mixed tree cover over Britain. It was thought that the pattern of life was much the same from about 5,500 B.C. to the second millenium B.C., when mesolithic or middle stone age hunters and fishermen sought their food in the less heavily wooded areas on the fringe of pine and birch forests, and fished the waters bordering the sand dunes of what is now called Breckland in north-west Suffolk and adjoining Norfolk. However, recently layers of charcoal have been found in sediments dated to the later part of the period which raises the question whether these were formed by forest fires caused by lightning or could they indicate the beginnings already of the clearing of patches of woodland for shifting agriculture? In any case, it is at levels corresponding to the period when neolithic people some 5,000 years ago began to practise agriculture in the easily worked soils of Breckland, that there appears a sudden significant rise in the pollen of heather, herbs and cereal as well as layers of charcoal and a decrease in tree pollen. It is thought that this increase in heather and other non-tree pollen can be explained by the fact that neolithic farmers cut down and burnt trees to create small clearances in which to sow seeds and graze herds, and as C. H. Gimingham (1975) has pointed out, where trees are excluded and soil becomes poor and acid, in the right climate heath can be expected to develop.

A great advance in our knowledge of prehistoric vegetation has been made possible by the new scientific tools of radio-carbon dating and pollen analysis. The pollen grains of flowering plants and spores of fern and moss which have lain unchanged for millions of years in layers of peat and other deposits can be identified with the plant to which they belong and, by working out the proportion of each species in layers of soil of known archaeological date, it has been possible to build a likely picture of the earth's vegetation at a given time. According to Gimingham, wherever a core has been taken from a peat bog on or near heathland and the pollen analysed, it has provided evidence that forest preceded heathland on lowland sites.

In *History of the British Flora*, H. Godwin (1975) describes how he arrived at the now generally accepted conclusion as to the origin of most lowland heathland when analysing pollen preserved in the deep mud of Hockham Mere in Breckland near Thetford, Norfolk. The fact that in certain deep layers of mud the pollen was almost entirely tree pollen led him to believe that Britain must have had a more or less complete woodland cover when the neolithic farmers first arrived here. However, not far from this level in the mud, there was a sharp fall in tree pollen, particularly elm *Ulmus glabra* with a big, simultaneous rise in the pollen of grass Graminae, ribwort plaintain *Plantago lanceolata* and sorrel *Rumex* spp. This pollen pattern pointing to elm decline and increase in non-tree pollen occurs in parts of north-west Europe around 3000 B.C.—the period when radio-carbon dates for neolithic cultural material suggests there were big movements of people to these parts. Bearing in mind that Breckland was as densely populated in neolithic times as any part of Britain, and that the great flint mines of Grimes Graves lie only a few miles away, Godwin became convinced that the origin of the present heath communities of Breckland was connected with forest clearances by neolithic agriculturalists.

In the same way we are recently learning more about the stone age environment and the origins of our heaths by the analysis of pollen in fossil soils that for thousands of years have lain undisturbed under burial mounds and other tumuli of neolithic construction. In the early 1970s, prior to Norfolk County Council building a school on open wasteland known as Brome Heath in Ditchingham, Norfolk, excavations were carried out at the site of an ancient monument that might have been affected. This was an earth enclosure marking the existence of a neolithic settlement where there was evidence of a long period of occupation before the earthwork was built, dating back probably to the fourth millenium B.C.

Buried beneath the earth bank was a surface of fossilised soil of neolithic age which it was hoped would throw light on the nature of the environment before the mound was built. G. J. Wainwright (1972) describes how soil samples were taken from this ancient soil surface and analysed for pollen. From the results it appears the landscape before the enclosure was built was probably a clearing in hazel scrub covered with grasses, rib-wort plaintain, and such plants of open spaces as heather *Calluna vulgaris* and sheep's-bit scabious *Jasione montana* with only a sparse growth of trees, mainly alder *Alnus glutinosa* and hazel *Corylus avellana*. This soil contrasted with the good brown-earth of lower levels where high tree pollen was found. The soil immediately under the earth enclosure did not show signs of cultivation but was probably used for pasture, and a thin top layer of infertile sand suggested that over-grazing before the construction of the bank had created conditions typical of acid heathland, where previously there had been forest.

But how could our forefathers' farming practices have helped to turn woodland into heathland and alter the soil so that it suited the heather plant community? Winifred Pennington (1974) in *The History of British Vegetation* offers an explanation. Under deciduous woodlands containing oak, holly and hazel such as new stone age farmers would have found covering much of Britain, there develops an acid brown-earth in which there is little evidence of podsolisation (page 67), but removal of tree cover or even grazing the field layer can trigger off rapid changes in the soil.

These acid brown-earth soils are in a delicately balanced equilibrium. The trees create a strong upward transpiration of moisture and much of the food taken up by the roots is returned in solution to the subsoil through the leaves. The soil's store of nutrition is therefore continuously circulated both by the vegetation and the soil fauna in such a way as to maintain the equilibrium. Any interruption of this circulation by such action as the felling of trees, for instance, can have drastic consequences. It is thought that changes in the soil cover associated with the farming activities of early people and the

40

cropping of their flocks could have constituted such an interruption and helped to transform forest brown-earth to an acid podsol-type soil on which heather flourishes. By its own highly acid humus, heather itself plays a part in the process as organic acids from its plant remains dissolve in rainfall and speed up the leaching of salts from the mineral soil horizon.

It is generally thought that forest clearance only became significant from about 500 B.C. onwards when climate changes less favourable to tree growth coincided with the development of iron age technology. Such a period of technological advance arrived with the axes and heavy iron ploughs of the earliest Saxons who, judging from the treasure found in the Saxon burial ship on Sutton Heath, came from Sweden to settle in the Sandlings of south-east Suffolk where they were later known as the Sandlings People. In the sixth century these light lands formed the central core of the Wuffinga, the powerful East Anglian kings whose kingdom stretched between the estuary of the Alde and the Deben, with the royal seat at Rendlesham where today forestry and a bomber base have taken over the heathland. The Sandlings People were the ancestors of the Norsemen who later came to East Anglia and old maps and documents show that further forest destruction took place during this Viking period of settlement (780–1070), continued through mediaeval times and occurred periodically thereafter with varying intensity until the end of the nineteenth century when the extent of heathland probably reached its peak in Britain.

It must be stressed that while human activity has helped in the conversion of forest into heathland, as part of a plant succession absolutely vital to this process are certain conditions of climate and soil. In the later pliocene period, nearly two million years ago, long before people could have had an effect on the environment, the great cooling of the world's climate at the approach of the ice ages, accompanied by much moisture and consequent acidification of the world's soil, provided conditions unfavourable to tree growth but very favourable to the spread of some plants of the heather family and grasses which soon covered large areas of the earth's surface. Heath appears also to have been a successional stage from forest on the acid soil of at least two later interglacial stages of the pleistocene. Pollen assemblages, for instance for the most recent interglacial, show a progression from forest to dwarf shrub *Empetrum nigrum* heath, with no trace of any influence of (palaeolithic) man, a change which must be attributed to soil changes. Also, as already mentioned, today in northern Britain there are places like the Hebrides where there is no evidence that the heaths were any other than a *natural* progression from sparse forest to heath.

COMMON RIGHTS—A HERITAGE OF STRUGGLE

Sir William Blackstone in his *Commentaries on the Laws of England* first published 1765–69 writes that "Common is a profit which a man hath in the land of another so as to pasture beasts thereon, catch fish, dig turf, to cut wood or like." Over the centuries, the changing ownership of wild land or waste has resulted in a continuous tug of war at village level between the owner of the soil and the owner of these common rights.

There are virtually no written records until the ninth century but the picture these give of mediaeval society suggest the very early importance of the mark or waste. Folkrights over heath, marsh and wood were part of a very ancient pattern of agriculture and an important part of the village economy for here fruit could be gathered, pigs root for acorns and nuts, beasts could pasture and turf and furze could be cut for fuel.

However, the wording of a Norman statute known as Westminster II (1285) shows that already in the thirteenth century the peasants were suffering encroachment on their commonland and, as has happened all down the ages, under cover of darkness privately erected fences were pulled down with the tacit approval of the village community. The statute reads:

> And since it sometimes happens that one having the right to approve has raised a bank or hedge and some persons by night or some other time when they do not think their deed will be known have thrown down the bank or hedge and the men of the neighbouring townships are not willing to indict those guilty of such a deed, the township adjoining shall be distrained to raise the bank or hedge of their own expense and make good the loss.

By Norman times, just as the king already claimed all the forest, the lords of the manor were claiming all rights over the waste and the statute quoted above together with a law known as the Statute of Merton in 1236 gave legal right to the lord to enclose the common "as long as he left the commoners a sufficiency for their needs".

42

According to Stephen's *New Commentaries on the Laws of England* (1841–45) it is to these two doctrines that we owe "the legal doctrine that the soil of the waste is the lord's freehold and consequently all rights of the commoners are derived expressly and by implication from him". If the law had been put into practice, it would have meant the virtual end to ancient folkrights, but as A. L. Morton (1948) points out, the determination of the peasantry proved more than a match for the Norman lawyers and the law was never in practice fully implemented.

In theory as head of society the Norman king held the freehold of all the land which he parcelled out in large manors and granted to his French barons, the manor being imposed on the old township and becoming the new agricultural unit. Leistuna—or Legestona (1168) or Leeston (1179)—was taken from its Saxon overlord and presented to William Malet and his son, Robert. The manor which included Leiston, Aldringham and part·of Thorpe features in the Domesday Book Survey as one of the largest estates in the area, comprising some 1440 acres of land. The high number of freemen recorded for the township of Leiston (Leistuna), about equal in number to villeins and serfs combined, is among the highest for Suffolk where the proportion of freemen is greater than for the rest of Britain.

The entry for Leiston in the Little Domesday Book survey of 1086 mentions sheep, hives of bees, great beasts and swine—all likely to make use of the waste. The chief interest lies, perhaps, in the decrease in the number of swine as a possible indication of tree clearance and growth of heathland and pasture. In 1086 there was sufficient wood to support 200 swine whereas twenty years earlier in the time of the previous king, Edward the Confessor there had been sufficient for 500. During the same period plough teams at work in Leiston had decreased from 17 to $10\frac{1}{2}$ so provided the figures are correct, this suggests as H. C. Darby (1971) quoting R. Lennard (1945) points out, that the "wasting" of the woodland was not due to extra plough-up. However, Oliver Rackham (1976) reminds us that the fall in swine numbers could indicate a difference in management of woodland rather than a decrease in its extent, perhaps a change over to coppicing to provide smaller, more manageable trees for a variety of domestic uses but fewer mature oaks to produce acorns for pigs.

With the break up of the manorial system, which accompanied the decline of farming methods based on open field arable land and the common use of meadow and waste, individuals had more control over their farmland and farming system. Some consolidation and enclosure of the arable strips in the open field arable land started even in the twelfth century but this greatly increased in the fourteenth and fifteenth centuries. During the Tudor period and the following century, larger amounts of arable land were enclosed and this included most of the arable enclosure of the clay lands in central Suffolk (Tate 1957). In this conversion of ancient methods of husbandry to modern systems of individual farming based on separate land ownership and cultivation, much heath and common-land not under the plough was enclosed. This was taking place even in the Sandlings of Suffolk for the surveyor Norden in his elegant map of the estate of Sir Michael Stanhope 1601–2 between the Deben and Alde estuaries, marks some enclosed "heather grounds" at Tunstall and a large enclosed area of "late common" at Eyke, as well as enclosed arable.

While it is true that it was not until the eighteenth and nineteenth century that the main attack was made on villagers' common rights, already in 1523 Fitzherbert was writing in his *Book of Surveying*: "The lords have enclosed a great part of their waste ground and straightened the tenants of their commons therein", while two years before a popular ballad was lamenting the changes:

> Commons to close and keep
> poor folk for bred do cry and wepe
> Towns pulled down to pasture sheep
> This is the newgyse.

Against this in *The Discourse of the Common Weal of this realm of England*, now attributed to Sir Thomas Smith (c. 1549), the Knight puts the case for improved agriculture that private ownership and separate cultivation bring: "That which is possessed of manie in common is neglected by all, and experience shows that tenants in common be not so good husbandes as when every man has his part in several."

Thomas Tusser (1573) in his *Five Hundreth points of good husbandry* held up his own early enclosed county of Suffolk where he farmed as a shining example of improved agriculture following upon enclosure, and his poem A Comparison between Champion (open field) and Severall (enclosed and individually farmed) is much quoted. In it he contrasts Suffolk with neighbouring Essex mostly unenclosed:

> More plenty of mutton and beefe,
> Corn, butter, and cheese of the best,

More wealth anywhere, (to be briefe)
Where find ye? (Go search the coast)
There where enclosure is most.

It is sometimes argued today that the loss of common rights was relatively unimportant to the poorer classes because an increasing number of countrymen were becoming hired day labourers, fewer and fewer owned any beasts to pasture on the waste and most benefited from the improved agriculture in increased employment. But whatever the *actual* effect may have been there is plenty of evidence that in the minds of countrymen they felt they were being robbed of their heritage. One memorable struggle of villagers in the south-east corner of Suffolk shows that in the seventeenth century they reckoned their ancient rights to be of real economic value.

It took place in the village of Walberswick described in 1662 as a haven by the sea, where today is found one of the few remaining extensive stretches of coastal heathland in the county. Caroline Christie writes of what happened in her diary of 1911:

> The common is always delightful golden in Spring time, with the sweet cocoanut scented gorse . . . In July and August it is purple with heather and then comes the orange and yellow of bracken . . . The common lands comprise some 1,400 acres with the fens.

In 1662, the inhabitants of Walberswick regained (their rights) after 30 years, Sir Robert having withheld them. But two years after, great trouble arose in consequence of Thomas Palmer, Sir Robert's great farmer, feeding his sheep on the uplands and putting them into East Marsh, breaking down the townsmen's posts and rails and filling up their dyke. Sir Robert also set up a boarded house for men and dogs near Paul's Fen to drive out any cattle belonging to the men of Walberswick. A terrible fight took place in which four men lost their lives. The marsh from this circumstance was called Bloody Marsh. Again the common and marsh was possessed by Sir R. Brooke. His son succeeded him and there was a trial at which Judge Fessant eventually gave judgement in favour of the lord of the manor who held it for a life-time.

The original petition to the king from the commoners of Walberswick against Sir Robert Brooke, Knight, reads as follows:

The first petition 1637 . . . by your subjects to the number of 300 concerning the great wrongs done unto them by Sir Robert Brooke, as well as exacting excessive fines on admission of tenants 10 times as much as used to be paid. Debarring from such common of pasture for their cattle as they have time out of mind held and enjoyed and detaining them on certain marsh grounds which of right belong to them. Upon which commons and marsh grounds the said Robert Brooke caused a great farmhouse to be built and he has enclosed a part of said marshland and common and laid to such farm house and there places tenants of whom receiveth much rent—to the disinheritance of your majesty's poor subjects. . . The said Robert with sheep, beasts, conies so overcharges the rest of the common not enclosed that they break into the petitioners enclosed grounds and eat up and destroy their grass and corn, and what they sow and plant they cannot eat. For many years have groaned under the burden thereof and many of them have been constrained at great undervalue to sell up and leave the country.

Because of their extreme poverty and perverseness* they are unable of themselves to redress these intolerable wrongs. They refer all acquainted with the aforesaid oppressions—to make final end and restore everyman his rights.

The whole incident shows the value of heath and marsh to the seventeenth-century village community and the way enclosure, proceeding then in piecemeal fashion by private agreement and often without the backing of the law, in this case at least contributed to the final dispossession and emigration of some of the small peasantry. This is of interest because generally speaking Suffolk is thought to be a county where the early enclosures caused little unsettlement, a point of view that is confirmed by the fact that the 1536 and 1597 acts against enclosure and depopulation were not applied to the county. The Walberswick events may have had more lasting effect than appeared at the time for White's *Suffolk* of 1874 gives the common rights of Walberswick a rare mention. Describing Walberswick as an ancient village by the sea with 303 souls and 1960 acres of land, it records:

Upon 34 acres of land every householder has the right to turn one head of cattle. On 40 acres of salt marsh all parishioners have the right to turn what stock they choose, and the poor avail themselves of the privilege by feeding upon it great quantities of

* Perverseness is here used in Milton's sense of 'adversity'.

46

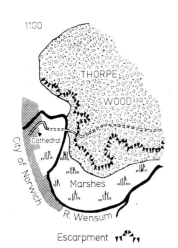

1100

THORPE WOOD

Cathedral

City of Norwich

Marshes

R. Wensum

Escarpment

RISE AND DECLINE OF AN ENGLISH COMMON. DEVELOPMENT OF MOUSE-HOLD HEATH, NORWICH.

1. AS DESCRIBED IN THE DOMESDAY BOOK AND OTHER TWELFTH-CENTURY DOCUMENTS WHEN THORPE WOOD COMPRISED THE WHOLE OF THE PRESENT MOUSEHOLD HEATH AND MORE, WITH WOOD FOR 1200 SWINE.

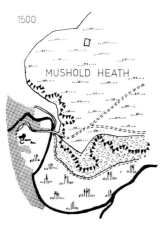

1500

MUSHOLD HEATH

Built-up

2. AS DEPICTED IN MAPS AND VIEWS OF THE SIXTEENTH CENTURY. THERE IS STILL MANAGED WOODLAND ON THE SOUTHERN SLOPES, BUT THE GRAVELLY PLATEAU HAS BECOME HEATHLAND WITH RIGHTS FOR SHEEP, CATTLE AND PIGS.

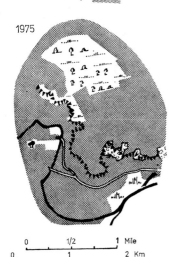

1975

0   1/2   1 Mile
0   1   2 Km

3. MOUSEHOLD HEATH TODAY. WHAT REMAINED AFTER IT HAD BEEN EATEN AWAY BY ENCLOSURE ACTS, AGRICULTURE AND SPECULATIVE BUILDING, TODAY SUFFERS FROM LACK OF GRAZING AND FIERCE FIRES SO THAT MOST OF THE HISTORIC HEATH HAS TURNED INTO SECONDARY OAK AND BIRCH WOODLAND.

A PAINTING BY JOHN SELL COTMAN SHOWS MOUSEHOLD HEATH IN 1809 AT A TRANSITIONAL STAGE. (SEE PLATE SECTION)

47

geese. A heath of 84½ acres in open pasture for all resident parishioners who also have the liberty of cutting furze, turf, ling. The tenant of Westwood Lodge (which replaced an ancient hall of the Lord of the manor) also has the right of turning sheep upon the heath.

It was Westwood Lodge Farm consisting of 3000 acres belonging then to Sir John Blois, that Arthur Young pointed to in his review of the agriculture of Suffolk as the finest farm in the county and an example of farms in the sand districts being much larger than those on the central boulder clay.

In their determined fight for common rights, the citizens of Walberswick were following an historic East Anglian tradition. In the same cause one hundred years before them, some twenty thousand Norfolk men led by Robert Kett were gathering on Mousehold Heath near Norwich intending to present the following petition to the king:

> We pray your grace that no lord shall common upon the commons . . . We pray that the freeholders and copyholders may take all profit of all commons . . . and that the lords shall not take profit on or profit off the same. That your lordship will take all liberties of leet into your hands wherebye all men may quietly enjoy their commons with all profit.

There is an entry in the Star Chamber Proceedings soon after the great 1548 rebellion concerning a certain Robert Brow who was accused by commoners of surcharging the common of Leiston, over stocking of commons by the lord's stock being one of the main grievances of the rebels. In self defence, he argued that his action was a legitimate reprisal for the conduct of the complainants on the common during "campanying" time! Campball was an old ball game which has been described as a kind of football fight of Saxon origin particularly popular in the countryside between the Orwell and the Alde, (A. S. campan: to fight.)

PARLIAMENTARY ENCLOSURES OF HEATH AND COMMONLAND

After the great outcry against ruthless Tudor enclosures, seventeenth-century enclosures were made largely by local exchange and agreement but in the eighteenth and nineteenth centuries Parliament itself put its whole weight behind enclosure in a series of acts relating both to open arable fields and to heaths, marshes, greens and other commons. In Suffolk, where the bulk of the enclosures on the boulder clay farms of High Suffolk had taken place much earlier, only about 96,104 acres were affected by parliamentary enclosure acts and these were mainly concerned with commons and heathland.

Arthur Young in his *General View* already quoted states that Suffolk must be reckoned one of the earliest enclosed counties yet in the last decade of the eighteenth century when he was writing, it is clear that a sizable amount of heathland remained unenclosed in the Sandlings for, while he points to whole tracts of the sandlands that have been converted from warren and sheepwalk into cultivated enclosure as part of the excellent husbandry he so much admires in this area, he also makes note of some customs which he sees as "impediments to the cultivation of the soil, these being rights of common and pasture which exceed the ordinary". He calculated that for the whole county, wastes comprehended under the term walk, warrens, common etc amounted to some 100,000 acres or an eighth of the whole and comments:

> Whoever has viewed the immense wastes that fill the whole country from Newmarket to Thetford . . . and which are found between Woodbridge and Orford, and thence way to Saxmundham, not to mention the various heaths that are scattered everywhere, must be convinced that their improvement for grass would enable the county to carry thousands more sheep.

His annotator disagrees however and argues that the heaths are so poor that they are best left undisturbed particularly since upon new broken up lands the lambs are subject to rickets. He refers to the rage for ploughing up and exclaims:

> Within a few miles of me, several heaths which were broken up and improved under skilful occupiers, about 30 years since, have within these last ten years been laid down again and reconverted to heathland . . . and it may be observed that the new heath will not be so good as the old one for more than 20 years to come.

While Young is much in favour of enclosure for the improvement it brought to agriculture ("Capital converted barren heathlands into smiling cornfields", he wrote in *The Farmer's Tour Through the East of England* 1771), he nevertheless observes and records the hardship enclosures of heaths and commons brought to the labouring poor. Several acts made generous provision for the poor. At Snettisham in Norfolk, Young describes how instead of cutting whin all over the open fields as previously, the poor were allotted 100 acres of common in one enclosure for cutting turf. At Barmingham in Suffolk, where there were previously 400 acres of wet commonland, after enclosure 20 acres with furze upon it was allotted to the poor for firing, while at Brandon in north-west Suffolk when 4,500 acres of open sheepwalk and rabbit warren was enclosed in 1807, 163 acres of sterile land at Lingheath Farm were let with a view to purchasing fuel with the proceeds for distribution to poor parishioners. (White's *History of Suffolk* 1874.) But in the report of a government *Inquiry into the Propriety of Applying Wastes to the Better Maintenance and Support of the Poor* (1801) Arthur Young is quoted as saying: "by nineteen enclosure acts out of twenty, the poor are injured, some grossly injured."

In contrast to his Norfolk survey, Arthur Young reviews few cases of enclosure in Suffolk in his *General View* of 1794 but in his account of Barton Mills village he shows clearly what could happen to smallholders with the enclosure of heathland:

> The division of the heath answered greatly, 300 acres of it have been broken up and are under corn etc. Corn much increased. Sheep halved, cows annihilated . . .

Under the heading *The Poor* he continues:

> There are 18 common right houses and 8 acres a year were assigned to each. The poor people of all did not suffer, but little owners of 20L and 30L a year suffered severely. A few such who by means of a common, summer fed 7, 8, or 10 cows and wintered then on the straw of their arable . . . found themselves allotments of 20 to 25 acres of good land which neither feed a team, nor half the cows (had they attempted to do it) which they had before. They were forced to sell immediately.

Cottage tenants with no proprietary rights do not appear in enclosure acts so no statistics are available, but in his review of Norfolk agriculture Young said that in many cases "poor people suffered who have turned cows and geese upon the common but have no rights". Evidently villagers who possessed no rights attached to property had often made use of commons. This seems to support A.

J. Peacock's belief that cottage labourers may have been the real victims of some enclosure acts. In *Bread or Blood* (1965) an account of the riots in East Anglia in 1815, he points out that the disadvantage to them was aggravated by having to buy fuel which previously they had obtained from the common at a time when turf at 7s a 1000 and coal at 40s a cauldron was very dear in relation to the current weekly wage of 11s.

Bitterness at the loss of "labour's rights" is heard in a poem by John Clare, described in *Poems descriptive of Rural Life and Scenes* published about this time (1821) as the work of "a Northamptonshire peasant":

> Now the sweet vision of my boyish hours
> Free as spring clouds and wild as forest flowers
> . Is faded all—a hope that blossomed free
> And hath been once as it no more shall be.
> Enclosure came and trampled on the grave
> Of labour's rights and left the poor a slave . . .
> Moors losing from the sight, far, smooth and blea
> Where swept the plover in its pleasure free
> Are banished now with heaths once wild and gay
> As poet's vision of life's early day.
> Like mighty giants of their limbs bereft,
> The skybound wastes in mingled garbs are left,
> Fence meeting fence in owner's little bounds,
> Of fields and meadows, large as garden grounds,
> In little parcels little minds to please
> With men and flocks imprisoned, ill at ease . . .

Here is that same sense of cultural loss already referred to, connected with some imaginary ancient freedom when "No fence of ownership crept in between".

ENCLOSED BY ACT BUT NOT IN DEED

It was by a parliamentary act drawn up five years before the East Anglian bread riots of 1815 that Leiston Common was given permission to enclose, at a time when the high corn prices during the Napoleonic Wars had been an enticement to farmers to plough up heathland. Three years before the end of the wars with France on May 10th 1810, an Act for Inclosing Commonland within both the parishes of Leiston and Theberton received the royal assent but it was not until fourteen years later that the enclosure was finally awarded by which time Britain was in the middle of a post-war crisis and a slump in prices,

The act covered 200 acres of "commonable and wastelands of Leiston Wet Common, Leiston Dry Common, Whynter's Heath and the Valley lands . . . to the soil of which Joshua, Lord Huntingfield lays claim" which together with fen, bog, dry commons and greens of the neighbouring parish of Theberton made about 450 acres. Also involved were "divers other persons, owners and proprieters of certain ancient commonable messuages and cottages . . . and lands thereto respectively belonging."

Presumably the required procedure was followed of calling a meeting of landowners concerned to draw up a petition to enclose, after which commissioners were appointed. These commissioners were then authorised to fix notices on the doors of the churches of Leiston and Theberton "upon some Sunday immediately after Divine Service" giving notice of the act and the hearing of objectors. The wording of the notice gives some idea of the commoner's rights, suspended in this case during the delay of fourteen years at least.

It shall not be lawful from and after the passing of this Act until the execution of the award—to cut, dig, pare, grave, flay or carry away Rushes, Turf, Whin or Furze—without licence or consent of the said commissioners in writing (fine not exceeding £5).

The commissioners in question were members of the county gentry from neighbouring villages—Will Shulden of Marlesford in the County of Suffolk, Esquire; Robert Crabtree, Gentleman, and Robert Boyden, Gentleman of Northcove in the same county. After receiving objections in writing they "must then hear claims with all due speed." The act in its preamble argues that "whereas the said Common, Fens Marshes and other commonable and waste lands in their present uncultivatable state yield little profit but if divided and allotted among the several persons having right of common thereon, might be inclosed, cultivated and enclosed."

As often happened during enclosures in the interest of consolidation, private exchange of land took place among the big landowners who also bought out holdings previously owned by small holders who may have felt bound to sell because the high cost of enclosure would have been beyond their means. In practice, however, according to W. E. Tate (1957), Leiston is one of the relatively few examples of a common that was never actually physically enclosed. This may have been due to the fact that by the time the long delayed award was finalised in 1824, the catastrophic fall in the price of corn had brought many light land farmers to the verge of ruin.

Of course, conversion of heathland into cultivated land could and did take place without enclosure. Tenant farmer Edward Dewy rented 113 acres in the Leiston Common area from Lord Huntingfield in the rather more prosperous times of 1841 and details of his holding in the tithe survey of that year make interesting reading because his farm illustrates the way land was parcelled out on the light lands of coastal Suffolk. Each farmer usually received as part of his holding a portion of marsh, shingle and dry heath as well as the main arable area. Dewy's farm included four fields of arable named Twenty Seven Acre's Walk, Twenty Acre Walk and Eel Hole Walk (the term "walk" usually indicates that such land was formerly sheep walk or heath); three pastures named Long Meadow, Rushy Bottom, and Marshe with Reed Yards, together with two areas of Whin and one described as "benthill, dunes and shingle".

HEATHS AS MEETING PLACES

Greens, heathy commons and other open spaces have from early times provided a traditional common ground for social and political gatherings of local people and some, like the various "Coldfair Greens" one comes upon scattered about the country, are named after the great annual fairs that customarily took place upon them.

In the half century following the final enclosure award of Leiston Common, Leiston grew into a little industrial oasis round Garrett's Ironworks which by 1866 was employing 600–700 in ten acres of workshops and making an international reputation with its combined threshing, drawing and strawdressing machines. It seems possible therefore, that the area immediately round the works was less severely affected than other parts of rural Suffolk by the depression that hit the thirties and forties with the rapid fall in corn prices in the decades following the Napoleonic wars. However, the County of Suffolk as a whole, according to John Glyde Junior, editor of the *Suffolk Garland* (1866) had been "for a great many years notorious for the extent of its pauperism, and prior to the Poor Law Amendment Act . . . it was more deeply pauperised than any other county in England."

Central to the discontent that gave rise to the Chartist Movement in the middle of the nineteenth century was the high rural unemployment and low wages, aggravated by the introduction of the new threshing machinery which was seen as a threat to jobs. In mid-nineteenth century rural Suffolk, outside Ipswich it was only near such pockets of industry as Garrett's of Leiston and Smith's of Peasenhall that the movement put down any roots. In this area as in bygone times, the greens and commons became the gathering places for the great crowds who came from miles around to hear Chartist speakers. On Boxing Day 1838, five to six thousand people met at Bigsby's Corner near Saxmundham where a procession was formed to march to Coldfair Green in the countryside outside Leiston. Asa Briggs in *Chartist Studies* (1965) quotes an enthusiastic reporter in the *Essex Mercury* of January 1st 1839:

> The day was delightfully fine. The sun shone forth in all its majesty and topped the hills with gold while all nature looked serene. The scene was indeed an exhilarating one though there was no music to enlighten [sic] it, save a thousand voices.

Eight years earlier Rushmere Heath near Ipswich had been the meeting place to which rural workers had been summoned to discuss labourers' wages and later in the century it was the new trade unionism that drew the crowds to the commons, greens and open spaces in the years leading up to the great lock-out organised by the Essex and Suffolk Farmers' Defence Association in 1873–74 when a thousand union men were dismissed (Thirsk and Imray 1958). In keeping with the close links between Methodism and early rural trade unionism, Joseph Arch (1826–1919), founder of the agricultural workers' union, spoke in the Methodist Chapel at Coldfair Green in East Suffolk, today a gorse-covered open space vulnerable to fire and for many years the site of an ancient Christmas fair. The biographer of Leiston born Bill Andrews (1870–1951) relates how the young Andrews, who himself was to become internationally famous as a workers' leader, went to this meeting and heard Arch "harangue the farmworkers" on the green.

PEOPLE AND HEATHLAND TODAY

In the years of economic depression at the end of the last century and during the first decades of the present, much heathland that had been under the plough in the prosperous years of agriculture was allowed to revert to heath. With the end of traditional sheep grazing husbandry in the 1920s, changes took place in land use. Many acres of heath and moorland have been turned over to pheasant and grouse rearing. On the commons of Leiston-cum-Sizewell, for instance, the local squire, a large landowner, planted copses of pine trees in the early twenties to provide work for the unemployed and shelter for his game birds. It is probably true that, as in many cases, we owe the continued existence of this heathland habitat, depleted as it is, to the pheasants and their needs. In this particular case, it is fortunate that the squire's family included a naturalist of some distinction F. M. Ogilvie who has left a valuable record of bird life in the area.

A recent very positive development with the growth of nature conservation has been the renting by the same family after the second world war of acres of coastal heath and marsh to the Society for the Protection of Birds. Together with sand dunes, reed beds, shallow brackish pools, woods and arable land, these heaths form part of the varied habitat of the Minsmere Bird Sanctuary—a designated Site of Special Scientific Interest which in 1976 was purchased by the society.

For over thirty years the reserve has managed to co-exist well with the neighbouring nuclear power station, oil slicks, common fires and other ills that life is heir to. Management has been imaginative in many ways—not least in its public relations with the siting of public hides on the seashore perimeter of the reserve which from the beginning prevented local people from feeling excluded and gained their friendly interest. Another excellent example of good relations is the annual opening of Minsmere to school parties when wardens put themselves at the children's disposal. On a sunny June morning, the peace of its heaths and lagoons quietly absorbs the hushed excited groups of school children clustering round their guides, drinking in the information and the wonder of the place. The future of conservation may depend to a great extent on such good ideas.

This brief survey of the inter-relationship of people and heathland through the centuries as exemplified by one specific familiar locality must end unfortunately on a gloomier note. As far as the general may be illustrated in the particular, the following figures for the heaths of Suffolk's eastern coastal plain suggest the speed at which our British lowland heathland is vanishing.

According to the last detailed estimate made by the County Planning Department, in 1920 East Suffolk, which was then a separate county, contained about 95 square kilometres of heath. In 1932 this had dwindled to 77 square kms, in 1954 to 34 square kms and in 1968 to 26 square kms. That is to say, between 1920 and 1968 73% of the heathland had been lost. After 1968 the rate of loss slowed down with the virtual halt to afforestation expansion but arable still continued to nibble so that by 1973 the total left was probably nearer 22 square kms.

Where has this heathland vanished? From 1930–68, of the 69 square kilometres lost, some 48% went to arable, 43% to forestry, 9% to airfields. Of the 26 square kilometres of heathland remaining in 1968, rather less than half was in some way protected by being registered as either commonland, National Nature Reserve or National Trust land. 21% was registered as a Site of Special Scientific Interest, which means that it is important for wild life but not protected, and 33% was quite unprotected.

Today, according to John Shackles, Assistant Regional Officer of the Nature Conservancy Council, with the exception of Blaxhall Heath nearly all the best heaths are now scheduled with about 4200 acres (1700 hectares) within Sites of Special Scientific Interest. (It is difficult to give a more exact acreage as the heathland components of these sites are not computed separately.)

However, in recent years, the loss of heathland has continued with housing development taking place at Martlesham and the plough-up of some thirty acres of Snape Warren. The quality of the remaining heaths varies considerably and far the most serious threat is the invasion of scrub in the absence of grazing. Tansley (1911) foresaw that in the present world climate if left to themselves these lowland heaths would revert to dry oak and birch woodland and this is now happening in many places. Of the various other causes contributing to the deterioration of the Sandling heaths since 1920 the most important are perhaps the following: the invasion of conifers; increasing pressure to use the commons for recreation with the spread of the motor car, resulting in picnic areas and eroding patches, fires and dumping; and the use of areas for military air bases and power stations with the accompanying pylons.

If this is typical of what is happening to our heaths and commons on a national scale the speed-up of registration as urged by the Common Preservation Society is an essential first step. On a personal level, the facts and figures presented can only increase the resolve to look more carefully and with more appreciative eyes at the heathland that remains—The rest of this book is an attempt, with expert help, to do just this.

Part II

A LIVING HEATH
A study area
on the Suffolk Sandlings.

SOILS OF THE SANDLING HEATHS

Walking along the beaches that flank the coastal heaths of south-east Suffolk one often finds lumps of a partially hardened fossiliferous sandy "rock", varying in colour from whitish yellow to rust red. This is the Crag—a local name for any sort of shelly sand—that underlies many of the Sandling heaths. Scientifically, Crag is the term used to describe a fascinating series of marine deposits laid down at different times over a period probably stretching from about five million to half a million years ago, after great earth movements had brought about changes in the relative height of sea and land, causing the sea periodically to flood this region of East Anglia so that over immense periods of time the region was sometimes above, sometimes below sea water.

For some thirty-five million years, almost the whole of north-west Europe was submerged under what later came to be called the Chalk Sea, when the skeletons of myriads of microscopic plants and animals that lived in its waters coalesced to form the chalk that underlies East Anglia.

In the Sandlings we are more familiar with the London Clay—the bluish-grey clay that is exposed beneath the Crag at Bawdsey and Felixstowe, in places in the Deben Valley below Woodbridge, and along the Orwell. When the Chalk Sea subsided some sixty-five million years ago and mammals began to take the place of giant reptiles, the London Clay sediment was laid down in a sub-tropical sea. Later the whole area underwent a gentle folding and tilting toward the North Sea. While the chalk was heaved up probably above sea level in the west, the east of Suffolk found itself once more under a warm sea. This was the sea in which the oldest Crag, the Coralline Crag, was formed in the pliocene age some four to five million years ago.

From Butley Creek northwards to Aldeburgh and the Gedgrave marshes in Suffolk lies the main extent of the Coralline Crag with an out-crop under the sea at Sizewell bay. It is a light whitish yellow and full of the delicate branching fronds of fossil polyzoa—colonies of coral-like creatures whose shells accumulated in sandbanks submerged in a shallow Crag sea. The large number of shells similar to those found in warm mediterranean waters suggest that it was a warm sea with free communication to the south, and recent new techniques including pollen analysis have confirmed this belief. Later, when earth movements opened the area to colder northern seas with the approach of the ice ages, a gradual change in the shells appears. This is reflected in the increase in northern species found in the later Crag deposits of the early pleistocene period—the Red Crag and the Norwich Crag. (See Spencer 1979 and Chatwin 1954.)

A sea that rose far above the present North Sea level formed the Red Crag, so-called because layers of oxidised sandstone stain it a typical rusty red. It is the second oldest of the Crag soils and underlies the heaths from Walton-on-the-Naze to Aldeburgh in Suffolk where it meets the much older Coralline. As a rule it rests on London Clay and in places in south-east Suffolk near Butley and Woodbridge it is over twelve metres thick.

Quite apart from its geological interest, the Red Crag is scenically very striking where it is exposed on the heaths and cliffs of the Sandlings. Nowhere is this more so than at Butley where there is a large exposed stretch of Red Crag in a pit on rough heathland facing Butley Creek. It is seen at its best on a late summer evening when the low rays of the sun setting over the river intensify the glowing colour of the sand and throw into relief the holes of all the numerous inhabitants who each have their particular niche in the Crag community—from sand martins and stock doves to vast colonies of rabbits.

This Butley Red Crag has been described as consisting predominantly of iron-stained quartz grains and shell fragments. Marine mollusca, including the distinctive left handed whelk, *Neptunia contraria* make up the bulk of the fossils but certain rare freshwater shells (*Planorbis*) have been found in the upper layers. Geologists deduce from the sedimentary structure of this Red Crag, and the type of fossils found in it, that much of it was formed as underwater dunes in a shallow coastal environment early in the Pleistocene period of geological time some one and a half to two million years ago.

From the Leiston area stretching north some forty miles, the character and colour of the Crag changes from the marine deposits of the Coralline, and the current-bedded sand banks of the Red Crag to a mixture of sand, shelly sands, pebbly gravels and laminated clays that make up the most recent Norwich Crag. Study of pollen and the fossils of minute creatures contained in the crag by West and Norton (1974) and others suggest it was formed in one of the interglacial periods of Pleistocene times from deposits that collected in a shallow sea near the estuary of a flooding river (probably the forerunner of the Rhine) at a time when there was still a land bridge between England and the continent. It is one of the so-called Icenian Crags and its base in places lies nearly fifty metres below the present sea level.

FLOWERS OF THE CRAG PITS

The shelly Crag soils are rich in minerals, particularly calcium, iron and phosphates and in the dry sunny climate of the Sandlings provide the right environment for certain species of plants that add variety to the more typical heathland plant community. This fact was observed by the poet George Crabbe who lived as a child near pits of the Coralline Crag in the Aldeburgh district. At the age of thirteen, Crabbe's interest in botany was stimulated by a copy of Hudson's *Flora Anglica* given him by a colonel in the Norfolk militia stationed in the area and while earning his living as an apothecary in Aldeburgh he evidently still found delight in botanising in these pits of the heaths and waste places for he wrote:

> Our busy streets and sylvan walks between
> Fen, marshes, bog and heath all intervene;
> Here the pits of crag, with spongy, plashing base,
> To some enrich the uncultivated space
> For there are blossoms rare . . .

In 1889, fifty-seven years after the death of Crabbe, W. M. Hind in his *Flora of Suffolk* pays tribute to the poet's contribution to the botany of East Suffolk and mentions that he added forty new species to the botany of the county. Among them were small annuals and other plants of our sandy Crag soils: the tiny white petalled shepherd's cress *Teesdalia nudicaulis* which Crabbe found at Leiston, Aldeburgh and Orford; two other small annuals—suffocated clover *Trifolium suffocatum* growing near the sea at Aldeburgh and Dunwich, and the pale green little mouse-ear *Cerastium semidecandrum* at Sizewell in 1805; also sickle medick *Medicago falcata* with its sickle shaped seed pods at Sudbourne and Orford.

SHEPHERD'S CRESS
TEESDALIA NUDICAULIS

Today, walking along the heaths of the

VIPER'S BUGLOSS
ECHIUM VULGARE

Thorpeness-Sizewell-Leiston Common stretch of the Heritage Coast as George Crabbe must have done many times, one notices the tawny colouring of the heath broken here and there by the vivid blue and purple clumps of viper's bugloss *Echium vulgare* and the softly subdued maroon of the hounds-tongue *Cynoglossum officinale*. F. W. Simpson (1965) who made a study of the flora of the Coralline and Red Crags of East Suffolk, mentions these plants as typical of the open country, waste places and disturbed ground around the pits along with such other plants as wild mignonette *Reseda lutea*: spring vetch *Vicia lathyroides*: common storksbill *Erodium cicutarium*: birdsfoot *Ornithopus perpusillus* and sand sedge *Carex arenaria*, all of which occur on these commons.

IMPROVING THE SANDLING HEATHS

Some heavily mineralised phosphatic fossils found in the basement bed of the Red Crag are derived from much older rocks which were broken up by the force and weight of the high waters of the Red Crag Sea. Among these are coprolites—some of which are thought to be the fossil droppings of animals that happened to be preserved in these rocks.

It was J. H. Henslow a former president of Ipswich Museum who first realised the value to agriculture of these and other phosphatic nodules found in the crag and it is his researches that led to the development of the young fertiliser industry in East Suffolk during the nineteenth century.

Much earlier, in 1771, Arthur Young describes how he saw Crag being used to improve the Sandling heaths when he made a farming tour of East Anglia:

Crag is a singular body of cockle and other shells, found in great masses in various parts of the county from Dunwich quite down to the River Orwell . . . I have seen pits of it from which great quantities have been taken to the depth of 15–20 feet for improving heaths. It is both red and white, but generally red and the shells so broken as to resemble sand.

J. Kirby in 1735 had already noticed its qualities:

Where it (the soil) seems in a manner barren, it is fit for improvement by Chalk, Clay and Crag, which last is found by experience to be preferable to the other two, and may be cheaper. The Heath Part, commonly used for sheep walks, might contain about a third of the Sandlands, before the Discovery of Crag; but many hundred acres are now converted into good Arable land by excellent manure.

William and Hugh Rainbird (1849) on the other hand, writing a century later, were less sanguine about lasting reclamation of many heaths, believing that in all probability they would "remain as they are—mere sand encumbered with furz and fit for nothing but rabbits and sheep-walk".

However when the Ogilvie family came to East Suffolk in the middle of the last century and bought some six and a half thousand acres in the Sandlings, for the most part marsh and heathland stretching from Thorpeness to Minsmere, the estate papers show one of Mrs Margaret Ogilvie's activities when establishing farmsteads and planting trees was the digging of crag and gravel pits with a view to improving the light lands of the heaths. (I am indebted for this information to Mr George Cook, secretary accountant to the Ogilvie Estate for fifty-two years.)

That the value of Crag was appreciated still earlier in the century is demonstrated by a special clause in the Enclosure Act for Leiston Common of 1810. This clause set aside

that piece of land of the said commons and wastelands . . . containing 2 roods and 25 perches in the said parish of Leiston in the said strip of the same common . . . for a public land and gravel pit for the use of the proprieters of land and estates within the said Parish of Leiston and their tenants . . . for the improvements of their lands and grounds and the formation and repairs of the

Roads belonging to the said Parish. (This pit is now overgrown and the town's right of access surrendered by the Urban District Council)

PEBBLY BEDS AND SANDY GRAVELS

Laid down just before the Ice Age and associated in geological time with the later Norwich Crag era are the large rounded flint and quartz pebbles of the Dunwich heaths in the Minsmere region of the Sandlings. In worked pits near Westleton village one can see very clearly how the graded pebbles have been laid down in beds of white sand gently sloping towards the south-east, with the lowest beds made up entirely of pebbles. These are known as the Westleton Beds and are thought by some to be relics of ancient buried beaches. Where according to Hey (1967) they occur ten miles north to south and nine miles east to west of Westleton they give rise to some of the best stretches of heather *Calluna* covered heaths in the Sandlings and the rolling knolls and valleys typical of the Westleton bed terrain give it a character all its own.

Although the Crag soils of East Suffolk and Norfolk are of special interest since they do not occur outside East Anglia, they lie immediately below the heaths only in certain places. Elsewhere they are covered by later sand deposits. South of Aldeburgh, for instance, in the area of Hollesley Common, the Red Crag is covered by micaceous sandy soils that form part of a series known as the Kesgrave Sands and Gravels whose stratification suggests to geologists that they were carried by a north-eastward flowing river, possibly an early river Thames when the earth was freezing up during a cold stage over half a million years ago.

Towards Corton and Lowestoft the top soil is formed of glacial outwash sands and gravels associated with the ice sheet that covered East Anglia about 300,000 years ago in the full glacial period of the Anglian stage of the Great Ice Age. This glaciation took the form of two great lobes of ice that brought with them material dragged up from the floors over which they had travelled. One prong of the glacier approached East Anglia from the west, bringing in its wake a load of boulders, chalky pebbles, sand and mud which it plastered over central Suffolk in the form of Boulder Clay. When the ice retreated, it gave rise to floods of melt water, washing out the lighter sands and gravels and fanning them out over the land surface. These include the Corton Sand Beds near Lowestoft that underlie the heaths of which W. M. Hind wrote at the end of the last century in his *Flora of Suffolk:*

> A more or less broken belt of heath and furze extends near the coast from Lowestoft to the Orwell where *Calluna, Erica tetralix* and *cinerea* (heather, cross-leaved heath and bell-heather) grow in more or less abundance. Occasionally it is dissected by arable or marshy land . . .

So, laid down sediment upon sediment over millions of years or spread in the wake of glaciers, these are the soils that have given rise to the plant and animal community of the Sandling heathlands and influenced the working methods of those who farmed its light lands. It is however an intriguing thought that the opposite is also true: over the centuries the plants and animals and farming methods have had an equally profound effect on the soils of our heathlands.

BUILDING A HOME ON HEATHLAND

One of the rewards of living on a sandy common has been the opportunity to explore at one's back door the repercussions of people's own particular struggle for existence—their domestic and agricultural activity—on the heathland animal and plant community. Every action like a stone thrown in a pool, has effects spreading in ever widening circles. One cannot even throw out the washing-up water for long without changing the vegetation where it lands. Equally left to itself the natural vegetation of this lowland heath will change and revert to scrub and woodland unless certain types of activity that helped to bring it into existence are continued. Contradictory as it may sound, it is nevertheless true that within the continuity of the common, the only constant feature is change.

This small fragmented area of coastal heath is a patchwork of vegetation determined partly by the treatment its three hundred odd acres have or have not received from people and animals. Almost everything that is happening to heaths and commons in southern England has happened to Leiston Common and in its very ordinariness lies its interest. For a hundred years from the middle of the last century to the middle of this, the sheep-walks from here to Thorpeness were crossed by a one track local railway line. Built eight or nine years after the new threshing machines of Garrett's of Leiston had won acclaim in the Great Exhibition of 1851, the track was abandoned under the Beeching axe and has now reverted partially to heathland, leaving its memory behind in such field names as Fifteen Acre Rails and Stunnels. During World War II, the common was taken over as a battle school and its dwellers evicted. That too has passed. Now plants and allied insects find moisture and shelter in circular depressions that were once the foundations of army tents, and rabbits and brambles have taken possession of the old bomb craters. When that war ended, the national plough up policy was vigorously put into practice by the tenant farmer of neighbouring Crown Farm. Between 1946 and 1949 he ploughed up a total of 146 acres of the heaths and by adding fifteen tons of chalk to each acre of heathland, produced reasonable crops of barley, rye, kale and sugar beet. Nightjars nested in the heather that covered twenty-eight acres of the common bordering Lover's Lane, but with the plough-up of this last extensive area of heather *Calluna vulgaris* in the north-west corner of the common in 1948, they departed. Seaward, the coming of the nuclear power station engulfed much of Sizewell Warren but had the effect of preventing other housing and commercial development, and for several years now, black redstarts have built on its walls and red-backed shrikes on the perimeter of its grounds. Leiston and Sizewell commons today lie within the Suffolk Coastal Heaths Area of Outstanding Natural Beauty and the 34 miles of Suffolk Heritage Coast. Notwithstanding this, the building of a second and possibly a third nuclear power station at Sizewell is under discussion and at the time of writing seems likely to prove the focus for yet another form of popular struggle concerning the use of one-time open land.

THE SOIL PROFILE

Forty years ago, these events were part of an unknown future. In the spring of 1938 a wide stretch of heather still covered the common between the marshes to the north and the enclosed strip of grassy heath where a single bee-hive marked the place that was to be both work place and home. South and south-east the land sloped gently down to Sandy Lane, a track marked on the tithe map of 1841 as a public way which is today a bridle path cutting across the heath over Broom Covert to Sizewell beach half a mile away.

Arriving in a high west wind to view the site for a studio, the nature of the soil was vividly impressed upon the mind by the clouds of top soil blowing off the rye field opposite. (The west wind was still the prevalent wind, bringing rain and at its most turbulent, throwing the whole common into turmoil.) Awareness of soil structure deepened when boring began through some fifty feet for well water, an activity that revealed a soil profile not unlike that of old sand dunes. Beneath the grass leaves and stalks there was a very thin layer of raw humus made up of slowly decomposing plant remains. Beneath this again, a layer of roots of grass and some bell-heather rested on and often mixed with a layer of coarse whitish grey sand, nearly all the roots being in the top twenty cms of soil. Then came a band of slightly darker sand and below that as far as one could see raw unstructured yellow sand. There was little sign of the hard "moor pan" such as is often found at varying distances below heathland soils.

The highly acid humus produced by the remains of heather plants lies on the surface and rots very slowly in the comparatively cool oceanic climate in which plants of the heather family thrive. During this process humic acid is formed which when it is dissolved in the rain, seeps into layers of mineral material below and mobilises iron and aluminium oxides on the way. These too drain down through the porous soil and are often deposited below in bands streaked with the rusty red colour of iron. It is this iron deposit that sometimes forms so hard a pan that few plants can penetrate it. The bleached ashen grey appearance of the drained upper layers gives the name *podsol* to these acid heathland soils. (Podsol from Russian *pod* meaning under and *sol* meaning ash.)

We soon learnt when digging the foundations for the house, constructing the cesspit and laying out the garden area, that the soil varied considerably within quite short distances and that while plants were obviously affected by the soil, the soil in its turn was affected by the humus associated with the dominant plants. Since many heathland soils are derived from acidic siliceous rocks, we were expecting a high acidity but tests taken over an area of three to four acres gave an average pH value of 5 to 5.50 (pH value 7 indicates a neutral soil and the lower the figure falls below that the more acid the soil.)

These soil tests are based on the fact that soil contains, besides mineral grains, organic material mixed with varying amounts of clay. We need to know a lot more about how it works but this clay/humus complex has properties of weak acid and can absorb salts such as calcium. In soil derived for instance from chalk and containing many

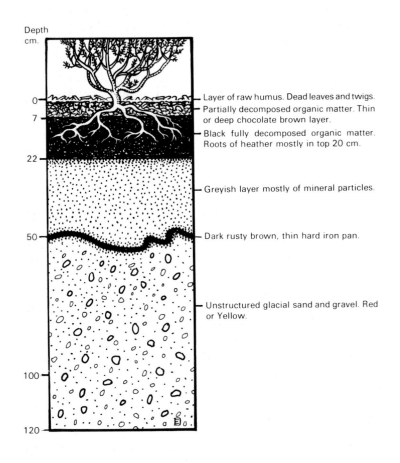

Depth
cm.

0 ——— Layer of raw humus. Dead leaves and twigs.
— Partially decomposed organic matter. Thin
7 — or deep chocolate brown layer.
— Black fully decomposed organic matter.
Roots of heather mostly in top 20 cm.
22 —

— Greyish layer mostly of mineral particles.

50 — — Dark rusty brown, thin hard iron pan.

— Unstructured glacial sand and gravel. Red
or Yellow.

100 —

120 —

DIAGRAM OF THE PROFILE OF A HEATH PODSOL

particles of calcium carbonate, the clay/humus mixture takes up as much calcium as it can hold so neutralising the acid. However, if as on most heathland soils, there is little calcium carbonate to absorb, there are not sufficient salts or bases to react with the acidic constituent—hydrogen—and an acid soil is the result.

Most plants have a pH range, a degree of acidity, beyond which they do not occur, one in which they grow but do not flourish and one in which they grow at their best, so soil testing can be revealing and an important guide to land treatment.

One of the reasons for the variation in soil came to light later when exploring the various pits scattered around Leiston Common, locally called "the shell pits". These are referred to in the Enclosure Act of 1810 as "public Sand, Gravel, Stone, Clay and Marl pits". The larger pits are no longer accessible but just before Sandy Lane leaves Broom Covert on its way to Sizewell, there is a small pit where the sand is so shelly that until the last few years it was the custom of local people to collect a pail of its contents for chicken grit to provide calcium for egg shells.

In the sheltered hollow of the pit before it became a tangle of undergrowth, on the sun-baked earth where rabbits had excavated their burrows innumerable fragments of fragile fossil shells lay weathering in the sun, scratched up among the upturned earth. Here half a mile from the sea, the children of the common came in search for whole shells sometimes bringing back among their treasures fossils of perfect little turret shells, whelks and cockles, tellins and periwinkles, small and easily crumbled but similar to those found on the beach today, including the right handed whelk *Neptuna antigua* of northern seas.

This fossiliferous shelly sand is the Norwich Crag that comes to the surface on Leiston and Sizewell Commons and extends northwards to the Norfolk Coast where it is known as Weybourne Crag. It was geologist H. E. P. Spencer when assistant to the curator of Ipswich Museum who in 1959 pointed to the possible interest of the pits on the commons and suggested that a number of overgrown disused pits might pay for re-opening.

It is characteristic of Norwich Crag soils that they vary considerably within quite short distances in the amount of shell they contain. Some patches are rich in fossils and others have none at all and can easily be mistaken for glacial sands. When excavations were being made on Sizewell Warren, for instance, for the first nuclear power station, the preliminary bore penetrated 129 feet without reaching the crag base and for most of this distance there was uniform Norwich Crag sand more or less devoid of fossils. In great contrast, at a short distance to the north near the site of a possible second power station, the soil of Sizewell Warren is rich in Norwich Crag fossils including the remains of pre-historic horse, deer and rhinoceros as well as the fossilised shells of a variety of sea and river creatures. It is from the less mineralised bones in the crag that we are able to build up a picture of now extinct mammals and other creatures that were roaming the coastal belt of Suffolk and Norfolk in Norwich Crag and Westleton Beds times: the Crag elephant *Archidiskodon meridionalis*, horse *Equus cabullus fossilis* and *E. robustus*, five or six species of deer including the giant deer *Megaceros verticornis*, a leopard *Felis pardoides*, rhinoceros, a giant gazelle *Gazella anglica*, various species of otter, beaver and large voles. (Spencer 1979)

This Sizewell Warren site has long aroused considerable interest especially since a pit survey, grant aided by the Nature Conservancy in 1953, found that the Norwich Crag here was unlike that of other Suffolk sites. David Long (1959), describes how a team of three collected fossils in the area over a period of four years and came to the conclusion that the Sizewell deposit would yield important evidence of the terrestrial and fresh water fauna of Norwich Crag should extensive excavations be carried out there. Twenty years later preparations for a second nuclear power station have added both urgency and opportunity for this further work much of which has been undertaken by R. A. D. Markham, geologist at the Ipswich Museum, who reports that the site continues to add information to our knowledge of the Norwich Crag fauna.

THE HEATHLAND VEGETATION MOSAIC

If soil and moisture levels are crucial in determining the pattern of heathland, so often is the biotic factor—its treatment by living things. In this respect the effect on the common's plant and animal community of our setting up home in its midst was in itself an excellent lesson in heathland ecology.

Basically our purpose was to construct an area from which we could become as self sufficient as possible and, in building a studio home, to carry out all activities with minimum disturbance to the heathland so as to maintain the character of the common. The aim was to keep all building to a level comparable with the clumps of gorse and elder bushes so that the long low building and its outhouses were dwarfed by the heath rather than the reverse.

Only in the fruit and vegetable garden did we attempt to transform the sandy soil, preferring elsewhere to grow such herbs and flowering plants as favoured the sunny climate and dry porous soil. Terraces with low retaining walls were built to conserve the moisture and in the vegetable plots it was necessary to add what the hungry soil lacked: nitrogen and phosphates and other plant foods, and above all organic matter. Year after year, bracken cut from the common was stacked as mulch on beds and around young fruit trees or composted with pony and chicken manure. Gradually earthworms which do not flourish in acid heathland soils began to appear and do their good work of pulling the mulch down into the earth and turning the yellow sand into reasonably fibrous loam. In this way a small section of the common was reclaimed from the wild to make a fertile garden on sand.

Along with other inhabitants when Leiston Common became a war-time battle school we were evacuated from the common, in our case nearer the sea behind the dunes of Sizewell. It was a few years later, after returning in 1947, that we acquired a strip of heath each side of the house to rear chickens and graze a pony. Two totally different treatments of the land on either side thirty years later has produced strikingly different results, making its own distinct contribution to the mosaic pattern of the heath.

A SMALL WILDERNESS

A strip of heathland flanking the garden to the west was once ploughed before it became for the next five years the feeding and scratching ground for about a hundred chicken. After that time it was left ungrazed except by rabbits and allowed to revert to scrub.

The combined effect of a single ploughing and the nitrogenous chicken droppings was to grow a splendid crop of nettles and encourage the growth of bracken. At first we looked upon this as an undiluted evil. Then as tree lupins *Lupinus arboreus* moved onto the disturbed ground adding their pale yellow to the deeper gold of gorse and broom, and filling June evenings with their sweet peppery scent,

71

we changed our attitude to this particular patch. Blackberries growing here were large and luscious. Instead of struggling with them why not encourage them and simply treat the nettles and the bracken as an annual crop to be harvested, scything round the bushes to make glades and paths between? Stacked, the bracken had many uses and nettles made rich compost.

Scything needed to be carefully timed, keeping in mind the insect and bird life that this small wilderness was attracting. The nightingale frequently breeds among the scrub of the common and this patch is one of its chosen nesting places while in recent years low nesting scrub warblers like the lesser whitethroats and spotted fly-catchers have also nested here—none of them birds of the more open heath.

One reason for not cutting the nettles until late in the year is that caterpillars of the peacock butterfly *Inachis io* in June and July feed in batches on the common stinging nettle *Urtica dioica*, as do the larvae of the small tortoiseshell *Aglais urticae* on the underside of the terminal leaves. Nettles, too, are included in the food plants of the larvae of some of the most splendid moths found on Leiston Common like the cream spot tiger *Arctia villica* and the burnished brass moth *Diachrysia chrysitis* with its green metallic colouring.

The long trailing shoots of brambles sometimes need to be cut back to prevent them tip-rooting. Curving in an arch they reach the ground and send out roots to form the nucleus of a new bush. But here again care in trimming is needed if some of our important wild pollinators are to have a chance to survive. There are several species of solitary wasps and small black bees that excavate the pith of dead, dry bramble branches to make linear nesting burrows, so it is important to leave a fair proportion of short lengths of dead blackberry stems. As Jean Henri Fabre wrote in *Bramble Bees and others*:

> The peasant as he trims his hedge, whose riotous tangle threatens to encroach on the road, cuts the trailing stem of the bramble a foot or two from the ground and leaves the rootstock which soon dries up. These bramble-stumps sheltered and protected by the thorny brushwood, are in great demand among a host of Hymenoptera who have families to settle. The stump when dry, offers an opportunity to anyone who knows how to make use of a hygenic dwelling, where there is no fear of damp from the sap; its soft and abundant pith lends itself to easy work . . .

Fabre listed nearly thirty species of bramble-dwelling *Hymenoptera* in the neighbourhood of his house.

This small wilderness is a conservation area for other creatures less happy on the drier open heath with its extremes of temperature. Rain water and algae collect in an earthenware crock sunk in the soil for the past thirty years, in the first place to provide chicken water. Here toads gather and on damp summer nights can be heard plopping about here and in the garden area. This water may also help to attract short-tailed voles *Microtus agrestis* whose runs are visible in the longer, coarser grasses that grow here. In Spring 1955 diary records mention a female vole caught by the children and brought into the house in a wire cage provided with apples and grass. This proved to be a pregnant female which by morning escaped, leaving behind three blunt-nosed young, pink and bare as sugar mice. Gillian K. Godfrey (1953) described how she analysed field voles' droppings on Rough Common in Wytham, Berkshire, and found that their food consisted mainly of grasses chewed in fine pieces, including sheep's fescue grass *Festuca ovina* and other coarser grasses found here in this wilderness area where their thicker cover supplies more moisture for the voles than on open heath. In some years there are great build-ups of vole populations and in other years there are inexplicable drops in numbers. Recent research has shown that, though food shortage does not necessarily precede a crash in numbers, vole growth and reproduction may be linked to the quality of spring forage dependent upon winter sunshine and amount of rainfall in April and May (P. N. Ferns 1979).

Field or short-tailed voles are the main diet of the short-eared owl, a regular winter visitor to Suffolk coastal marshes and heaths, where on rare occasions it has nested in rough grass and heather.

A SEA OF BRACKEN

In the late summer, like most of Leiston dry common today, this area is a sea of bracken and its pungent smell is one of the most characteristic scents of the heath, strongest when the earth warms up after rain. The bracken ferns make

SHORT-EARED OWL

a solid phalanx right up to the edges of the trodden grass paths where the packed air-tight earth temporarily deters them.

There is an old saying "Under bracken gold, under grass silver, but only copper under heath." Certainly the foresters watch for bracken *Pteridium aquilinum* when seeking the best soil for planting conifers on heathland. This is because bracken not only likes deep and moist soil but thrives only in relatively frost free areas. It is thought that the stolons in effect "cultivate" the soil and so help to diffuse oxygen in it. A single plant of bracken, it is estimated, may occupy up to half an acre of ground, with a truly enormous store of food in its underground rhizomes.

Except for frost, this plant has almost no enemies. It is scarcely grazed by animals; in early spring the young curled shoots are protected by last year's growth of dead fronds and its rhizomes are safe beneath the deep litter from heathland fires. Even the rocking of gale force winds on our heaths has advantages for this fern. The fronds bend and sway producing a funnel for the rain to run straight down where it is needed. Left uncut it forms its own deep bed on which it thrives while suffocating its rivals. All this makes it a formidable adversary here as on many heaths where heather *Calluna vulgaris* and bell heather *Erica cinerea* are playing a losing game.

However Nicholas Culpeper (1653) in his herbal has a good word to say for the bracken fern:

> The roots being bruised and bottled in oil, or hog's grease make a profitable ointment to heal wounds or pricks gotten in the flesh ... The Fern being burned, the smoke thereof drives away serpents, gnats and other noisesome creatures which in fenny countries do in the night time, trouble and molest people lying in their beds with their faces uncovered ...

(Recent medical research, however, has raised the suspicion that bracken may have carcinogenic properties.)

In the past, bracken has been kept in check to a certain extent by regular cutting for farming practices.

> Get home with thy brakes ere all Summer be gone,
> For teddered up cattle to sit down upon,
> To cover thy hovel, to brew and to bake,
> To lie in the bottom, where hovel ye make ...

So Thomas Tusser in the sixteenth century advised in his *Five hundreth points of good husbandry* and until quite recently his advice

was followed. Dick Chatten, retired cowman of Leiston Common, remembers how in the first three decades of this century in Suffolk:

> We used to use a lot of bracken for litter. We would cut it green and make it up into big heaps—bracken-cocks they called them. They would heat up green, go sear, and never get any water into them but keep dry until you went and carted them with horse and tumbril.

Where bracken is dominant, the only time other smaller plants have a chance to flourish is before the ferns develop their full canopy which is not until June or later. In early spring, where there are scattered open patches of sand among the stands of bracken, little winter annuals flower—the common whitlowgrass *Erophila verna*, the spring vetch *Vicia lathyroides* with its single blue flower, the pinkish yellow bird's-foot *Ornithopus perpusillus* and the early forget-me-not *Myosotis ramosissima*, all of them tiny plants which flower and ripen their fruits long before the bracken is any height.

STUDY AREA—GRAZED OPEN HEATH

A great contrast to the controlled wilderness of scrub to the west are three acres of heath to the east of the house, the result of quite different management. Throughout the time when lack of rabbit grazing through myxomatosis was affecting the rest of the common, this heath was hard grazed by a pony used to rough hill grazing. A cross between a Welsh and Shetland pony, she came to us with laminitis, an inflamation of the hoofs that can be aggravated by too rich grass and we were advised to keep her on as nearly bare ground as possible. Hence the hard grazing the piece of heath received when for a period of some ten years pony and rabbits had the enclosure to themselves, producing a strange hummocky lunar landscape, lichen and moss covered, with short thin grass and some bare earth. Today the result is an area of open heath approaching more nearly the sheep-grazed Sandling heaths of former times and for this reason providing an interesting study area.

Since the death of the pony some ten years ago, rabbits though reduced have been the all important animals here as elsewhere on the Sandling heaths, their activities helping to determine the pattern of vegetation and having an effect on what ecologists have called plant sociology—the relationships between plants in the heathland ecology.

PLANTS OF OPEN GRAZED HEATH

With the death of the pony and consequent decrease in grazing, the very free seeding bell heather made a rapid recovery. This is a plant that is vulnerable to grazing and stands up to it less well than some other members of the heather family. C. H. Gimingham found in experiments in Scotland that with mild grazing the growth of heather or ling *Calluna vulgaris* was enhanced but bell heather *Erica cinerea* with its more open branching tended gradually to die out. When grazing was further increased the ling too was damaged and under really heavy grazing, grasses took over. This was the position here at the height of pony and rabbit grazing.

The study area of the heath slopes south, is very well drained and sandy with little humus and near the sea—a situation that exactly suits bell heather which is found in similar situations in the Faroes, West Norway and north-west France. To conserve moisture the dark green leaves, arranged in whorls of three round the stem, are permanently so rolled back as to cover the hairy underside except for a small slit. At an early stage, the young plants resist dry conditions better than *Calluna vulgaris* seedlings by sending down a tap root for moisture. On a southern slope like this where summer heat is intense and grazing slight, bell heather can become the dominant plant as is happening here.

A diary entry for August 7th 1968 recorded three years after the

EIGHT HEATHLAND GRASSES—1, PURPLE MOOR-GRASS *MOLINIA CAERULEA.* 2, BROWN BENT *AGROSTIS CANINA.* 3. HEATH-GRASS *SIEGLINGIA DECUMBENS.* 4. SHEEP'S-FESCUE *FESTUCA OVINA.* 5. WAVY HAIR-GRASS *DESCHAMPSIA FLEXUOSA.* 6. EARLY HAIR-GRASS *AIRA PRAECOX.* 7. COMMON BENT *AGROSTIS TENUIS.* 8. BRISTLE BENT *AGROSTIS SETACEA.*

death of the pony and some twelve years after the arrival of myxomatosis:

> The purple bell heather spreading with mosses and much lichen. Besides bents and sheep's fescue *Agrostis/Festuca* and early hairgrass *Aira praecox*, sheep's sorrel *Rumex acetosella* tinges the whole surface red where it occurs over large areas along with moss and lichen on almost bare soil. In June among the red of sorrel everywhere there are white pools of heath bedstraw *Galium saxatile* in the hollows.

Today, the bell heather has spread spectacularly from north to south of the slope and will soon extend over the whole enclosure under slight rabbit grazing.

The woody evergreen shrubs of bell heather can grow up to eighteen inches but here their low prostrate growth is encouraged where the plants are nibbled into compact growth by rabbits. The bell-like flower heads are a mass of crimson-purple bloom from July onwards, turning a bronzed rust in autumn. These old flower heads seem to hang on all the winter, still showing even when the heath is clothed in snow in February. The plants are less resistant to winter cold, however, than heather according to P. Bannister (1965). And this was evident during the harsh winter of 1978–79 with consequent late flowering.

In the study area, heather *Calluna vulgaris* is absent except for one elderly bush that is dying and collapsing in the centre but regenerating on the cropped verges where the heather is short, closely packed and dome-shaped. This is because each time a long leading shoot is nibbled back by a rabbit, the structure of the plant enables it to replace this shoot with two or more new ones from a stack of short side shoots and axil buds. This cropped heather appears young and vigorous because all the plant's energy goes into the production of young shoots rather than into wood.

MOSS AND LICHEN—A MINIATURE LANDSCAPE

When the bell heather is not in bloom, from a distance the heath has a look of tawny uniformity but a close-up view reveals a varied and intricate ground layer pattern mainly of moss, lichen and fungi with some sparse grass—a miniature landscape less than half an inch in height.

Carpeting the area between the heather community are a variety of heathland mosses: "Variety upon variety, dark green and pale green;

78

mosses like little fir-trees, like plush, like malachite stars, like nothing on earth except moss," as Thomas Hardy described them.

The hair mosses are specially characteristic of dry acid heathland and two varieties at least grow on this particular sunny strip of coastal heath: *Polytrichum juniperinum* and *Polytrichum piliferum*. These are the mosses "like little fir-trees", dark bluish-green in colour with cushions of stiffly upright little stems each with its spirals of minute leaves. On patches of bare ground, carpeting it "like plush" is another moss with the Latin name (these mosses have no other) of *Ceratodon purpureus* with purplish-red stems to its spore capsules and spreading triangular leaves. Very widespread, also, are the loose tufts of the brilliant yellow-green *Dicranum scoparium* growing through and between the low spreading clumps of bell heather. Here, too, but less frequent is a graceful moss with the long awkward name of *Pseudoscleropodium purum*, common on chalk grassland but also found on heaths. Its light green leaves so closely overlap the stem as to give a high silvery sheen to the whole plant whose stems intertwine in a loose weft-like form.

A typical heathland moss—a variety, *ericetorum*, of *Hypnum cupressiforme* now renamed *Hypnum jutlandicum*—grows in pale whitish green wefts elsewhere on the common, but here rather surprisingly only in the decaying centre of the solitary heather bush *Calluna vulgaris* where the shady conditions of this mini-climate are similar to woodland. The leaves of its very regularly branched stems are curved typically to one side.

Moss spores (the equivalent of seeds in flowering plants) are contained in beautifully designed spore capsules, held aloft on delicate wiry stems and colouring as the spores ripen to orange, brown and deep red. Hardy's "malachite stars" were probably the starry red-gold rosettes that the male plants of some heathland mosses (like *Polytrichum juniperinum*) bear on their stem tips.

Both mosses and lichen can absorb water and dissolved food material over any part of the body. They have no true root and are only anchored to the ground by thread-like rhizoids.

The lichens on the close grazed heath of the study area are nearly all species of *Cladonia*, fruit bearing lichens with a tough outer rind growing directly from a "platform" in elongated stalks and strands. Their whitish green colouring picks out their presence among the deeper greens of the mosses. Particularly widespread are the delicate sponge-like masses of *Cladonia impexa*. *Impexa* means "combed out" and suggests the intricate branch growth, divided and subdivided at the tips. Often the heath is littered with "balls" of this lichen, light as spun candy, which have been pulled up by some bird and tossed around. We have not seen this in action, but guess it is either pheasants or partridge who are the culprits.

On January 12th 1979, in ground frozen hard by a thick hoar frost and night temperatures of $-9°$ C, another lichen typical of acid heath, *Cladonia floerkeana* was in fruit, its round red fruiting ascocarps showing up bright as new paint against the pale green of the plant. On these light soils *Cladonia coccifera* is also common, its stalked cup-shaped "podetia" or "trunks" bearing rather darker red fruit round the rim. Also found in this habitat is *Cladonia chlorophaea* with its cups scale-covered both inside and out, *Cladonia pityrea* with "antlers" on the brims of the cups and *Cladonia fimbriata* with brown fruiting ascocarps. These together with the leafy *Peltigera spuria*, and the grey *Hypogymnia physodes* that grows on the twigs of dying heather were collected from Leiston Common and identified by Peter Lambert of Norwich Museum, as representing a fairly typical selection of lichens from dry southern maritime heaths.

It is a pity that most lichens have no English names. They are among the most ancient of all life, going back to the very beginning of the vegetable kingdom with some colonies over two thousand years old and they are interesting in having an unusual biological make-up. Two entirely different organisms—a fungus and one of the algae—associate to form a lichen. The fungus cannot conduct the process of photosynthesis, so it makes up for this inability by incorporating within its own structure algae—green plants that can use sunlight, carbon dioxide and water to provide food enough for both. In their turn the algae are shielded against excessive sunlight and probably supplied with inorganic substances necessary for growth.

FUNGI ON HEATHLAND

Besides the ubiquitous fairy cakes *Hebeloma crustuliniforme*, the fungus that has established itself widely throughout the grazed study area is the false chanterelle *Hygrophoropsis aurantiaca*. Walking here on an autumn morning when the frost is on the ground, one sees its soft orange clusters scattered in little groups about an inch high among the lichen and mosses and sheep sorrel *Rumex acetosella*. It seems to stand up well to frost and persists late into December, looking very attractive in the frosty field.

Also typical of the bare peaty soil or mossy areas between clumps of heather are the fairy clubs *Clavaria argillacea* whose fruit body is club-shaped and, when fresh, can be a bright yellow ochre colour. Both this fungus and the false chanterelle are among the larger heathland fungi listed by Roy Watling (1973) of the Royal Botanical Gardens, Edinburgh, but it is among the more vegetated areas of the Sandling heaths that one must look for some others on the list. One of these is the dainty little horse-hair toadstool *Marasmius androsaceus* with its black horse-hair of a stem and smokey brown rose-tinged caps that rise in procession along the branches of dead and dying heather or on pine needles. There is also an almost invisibly fine mycorrhizal fungus that lives in the roots of heather plants *Calluna* and *Erica* in an association that probably improves the plant's ability to make use of dissolved nutrients in these "hungry" acid soils, but we still know very little about how this association works.

In freezing conditions of the last day of 1978, we found another very typical heathland fungus in fruit on Leiston Common. In Broom Covert where the bracken is annually rolled flat for the pheasant shoot, near a small oak *Quercus robur* growing through the sear bracken were slender toadstools with olive-brown caps and yellowish stems looking rather the worse for wear. The skin or pellicle was slimy and easy to separate and B. M. Spooner of the Royal Botanical Gardens, Kew confirmed that this was *Mycena epipterygia*.

CYCLE OF CHANGE

Watt (1955) and Gimingham (1969) and others have pointed out that the mosaic of vegetation patches on heathland is often controlled by the life history of the dominant plant in its pioneer, building, mature and degenerate phases leading to cycles of change in the heath vegetation as a whole. Where, for instance, the heather and bell-heather growth is at its most vigorous, little else grows. As it declines mosses, lichens and other bryophytes creep into the gaps left by decay. As the gaps widen and the open patches increase, flowering plants of heathy places begin to appear like wavy hair-grass *Deschampsia flexuosa*, the tall downy heath groundsel *Senecio sylvaticus*, and the trailing St John's wort *Hypericum humifusum* with its neat prostrate growth, pale green leaves and small yellow petals often edged on the underside with black dots. Heather seed germinates best on these acid open areas where full daylight reaches the ground and where there is a little raw humus to hold the damp, so here too will be found as time goes on, the heather seedlings which by their individual sequence of growth will in their turn help to determine the future mosaic of the heath.

81

PLANTS

1	Hare's-tail Cottongrass	*Eriophorum vaginatum*
2	Common Cottongrass	*Eriophorum angustifolium*
3	Bog Asphodel	*Narthecium ossifragum*
4	Common Butterwort	*Pinguicula vulgaris*
5	Marsh Gentian	*Gentiana pneumonanthe*
6	Round-leaved Sundew	*Drosera rotundifolia*
8	Heath Spotted orchid	*Dactylorhiza maculata*
9	Bog Pimpernel	*Anagallis tenella*
10	Bog Moss	*Sphagnum sp.*
11	Soft Rush	*Juncus effusus*
13	Cross-leaved Heath	*Erica tetralix*
14	Lousewort	*Pedicularis sylvatica*
15	Western Gorse	*Ulex gallii*
16	Tormentil	*Potentilla erecta*
17	Gorse	*Ulex europaeus*
20	Bell Heather	*Erica cinerea*
21	Petty Whin	*Genista anglica*
22	Bilberry	*Vaccinium myrtillus*
25	Sheep's Sorrel	*Rumex acetosella*
26	Heath Bedstraw	*Galium saxatile*
28	Dwarf Gorse	*Ulex minor*
29	(Fungus)	*Mycena epipterygia*
32	Bracken	*Pteridium aquilinum*
34	Heather	*Calluna vulgaris*
35	Horse-hair toadstool	*Marasmius androsaceus*
36	Heath Dog-violet	*Viola canina*
37	(Moss)	*Hypnum jutlandicum*
38	(Moss)	*Polytrichum juniperinum*
39	(Lichen)	*Cladonia impexa*
40	(Lichen)	*Cladonia floerkeana*
41	(Lichen)	*Cladonia coccifera*
42	Heath Milkwort	*Polygala serpyllifolia*
46	Dodder	*Cuscuta epithymum*
48	Trailing St John's-Wort	*Hypericum humifusum*
49	Biting Stonecrop	*Sedum acre*
50	Sheep's-bit	*Jasione montana*
51	Harebell	*Campanula rotundifolia*
54	Burnet Rose	*Rosa pimpinellifolia*
55	Broom	*Sarothamnus scoparius*

INSECTS

7	Keeled Orthetrum Dragonfly	*Orthetrum coerulescens* (male) WS 60mm
12	Bog Bush-cricket	*Metrioptera brachyptera* (female) BL 19mm excluding ovipositor
18	Gorse Shield Bug in autumn	*Piezodorus lituratus* BL 13mm
19	Green Hairstreak Butterfly	*Callophrys rubi* WS 30mm
23	Fox Moth Larva in October	*Macrothylacia rubi*
24	Small Copper Butterfly	*Lycaena phlaeas* (female) WS 30mm
27	Green Tiger Beetle	*Cicindela campestris* BL 15mm
30	Bumblebee Queen	*Bombus lucorum* BL 22mm
31	Fox Moth	*Macrothylacia rubi* (male) WS 48mm
33	Silver-studded Blue Butterfly	*Plebejus argus* (male) WS 27mm
43	Emperor Moth larva in July	*Saturnia pavonia*
44	Emperor Moth	*Saturnia pavonia* (male) WS 57mm
45	Heather Beetle	*Lochmaea suturalis* BL 6mm
47	Oil Beetle swollen with eggs	*Meloë proscarabaeus* (female) BL 36mm
52	Cinnabar Moth	*Tyria jacobaeae* (female) WS 38mm
53	Mottled Grasshopper	*Myrmeleotettix maculatus* (male) BL 15mm
56	Small Heath Butterfly	*Coenonympha pamphilus* (female) WS 33mm

BURROWS IN THE SANDY SOIL: WHO LIVES IN THIS HOLE?

No one walking over the southward facing open heath of the study area can fail to notice one of its most intriguing features: the number of holes bored in the sandy soil from nearly a foot across to the diameter of a small knitting needle. These are the homes of a great variety of mammals and insects which, although their number must be legion, manage to tap the heath's resources without exhausting them by each keeping to its own particular niche in the living pattern of the heath.

THE WILD RABBIT

First and foremost in size of hole and overall influence on the heath's natural and social history is perhaps the wild rabbit *Oryctolagus cuniculus*. By Tudor times, Sizewell and Leiston Commons seem to have formed part of a vast warren and such a warren was clearly a lucrative and desirable possession. Rev. Alfred Suckling in his *History and Antiquities of the County of Suffolk* (1848) writes that in the reign of Edward I "the Abbot of Leiston's rabbit-warren seems to have been very extensive, as it spread into the parishes and hamlets of Aldringham, Buxlow, Frieston, Theberton and Thorpe." In the year 1299 the Abbot of Leiston "impleaded" John de Leyston for trespassing upon his manor at Leiston and driving away his hares— the earliest prosecution for poaching except in the Royal Forests known to Suckling who recounts the incident. He further records how a lease was granted to Henry Stredfield of Leiston Coney-warren for twenty-one years at a high rental of "£20 per annum and a fine of £125 dated April 19th the 13th of Charles II 1673–44".

When Rev. W. B. Daniel was writing his *Rural Sports* (1812), the fur of the rabbit was worth twice the value of the carcass. In his remarks on the Game Laws, Daniel points out that though "The Lord of the Soil may make Burrows in a common and stock them with Rabbits", this did not entitle a commoner to take a rabbit from the common. Penalties for taking one at night were particularly severe. The sporting cleric quotes a Game Law of George III which said that "If any person shall into a Warren or Grounds in the night time and take or kill any coney against the will of the owner or occupier of such ground . . . he shall be transported for seven years, or suffer such other punishment, by whipping, fine or imprisonment, as the court shall award."

After half a lifetime sharing a heath with rabbits one develops a complicated relationship. On the heath, as opposed to farmland and gardens, rabbits in moderation are both attractive and beneficial. On still evenings they come out of their burrows to graze on the open heath, moving like shadows or sitting still as rounded stones in the moonlight. Even in snow and nine degrees of frost, footmarks show that they are around stripping twigs of their bark and gorse of its spines, but on days of high wind they seem to keep below. Perhaps the best time to watch them is in the early morning when whole families sit in front of their burrows grooming themselves in the first rays of the sun. It is very noticeable, as E. P. Farrow (1925) pointed out, how closely cropped is the vegetation round the burrows, leaving the ground almost bare except for sheep's sorrel. Raising their heads from time to time to take stock of the situation, they crop methodically with a scythe-like motion, grazing all that they can reach by turning their head in a semi-circle, before stepping on to the next patch. According to Rev. W. B. Daniel: "The young doe frequently kindles out of the warren. She scratches a burrow about 2 feet deep where she prepares a bed for her young composed of fur plucked from her own body and some blades of grass . . . Here she suckles and attends them for six weeks." More recent observation has shown that the wild doe suckles only at night, opening the earth-covering only once in 24 hours.

On the three acres of open heath of the study area, the rabbits are safe from gun and gas and, since they are our greatest allies in keeping the open character of the land and their numbers are not sufficient to prevent the spread of bell heather, here they are entirely welcome. A gardener's viewpoint however, is rather different and we sometimes complain that the rabbits have forced us to live inside a wire enclosure! When they began jumping four foot walls, it was a choice between putting up wire netting or growing only the plants they would not eat and we chose the former. The result was interesting ecologically, because the areas of natural grass within the garden that the rabbits could not reach soon became full of minute flowering wild plants growing often so low against the soil that the rotary mower missed them. Harebells *Campanula rotundifolia* that once grew on the common but possibly could not flower under constant nibbling, have now seeded themselves onto a lawn where each year in July their wiry stems and blue bell-like flowers densely cover some six square feet of grass. Just before the flowering and seeding period mowing ceases but it is likely however, that some small flowering plants, particularly those with the rosette of leaves near to the ground, would disappear altogether if there were neither grazing nor mowing to keep down taller plants and so give them access to sunshine.

As long as rabbits were abundant, the change in the appearance of our Sandling heaths following the disappearance of sheep grazing was only gradual. It was with the arrival of myxomatosis that the essential character began to change. I. Haslam (1954), Chief Officer of Pests of the East Suffolk Agricultural Executive Committee at the time, reported that the first outbreak in Suffolk occurred at Easton Bavents near Southwold and was noticed in Leiston on June 9th 1954, spreading to the whole county by November 23rd. Mortality was quicker and more widespread where there were big colonies since fleas were mainly responsible for its spread. By 1955 rabbits were very thin on the ground but small pockets had reappeared in almost every parish. They seemed, according to the report to be mostly "surface" rabbits very difficult to locate as "they appear almost scentless so that dogs which have hitherto been able to locate them with little difficulty are no longer able to find them."

Twenty-five years later, rabbits are again becoming fairly abundant but seedling gorse, bramble, birch and pine which got away unnibbled in their infancy are now strong plants safe from destruction and fast turning the heaths into pioneer woodland.

Because of these possible changes in the character of commons, warrens and lowland heaths following myxomatosis, the Nature Conservancy asked R. M. Lockley at Pembrokeshire Field Station to conduct research into the wild rabbit, from 1954–1959. An intriguing account of this study was published under the title *The Private Life of the Rabbit*. From a specially constructed underground warren Lockley was able to observe through glass many interesting facts concerning their social life. The doe, it seems, is the centre of what could be termed a maternal society and the buck and doe remain tied to each other for the rest of their lives by their territorial allegiance to a home.

In order of preference, the rabbit's menu on Leiston Common would be likely to be heather *Calluna vulgaris* and fine grasses *Agrostis/Festuca*, bell heather *Erica cinerea*, some flowering herbs like harebell *Campanula rotundifolia*, seedling gorse *Ulex europaeus* and *U. gallii*, seedling birch *Betula pendula*, and very low down on the list to be eaten mainly in circumstances of much hunger—nettles *Urtica dioica* and *U. urens*, bracken *Pteridium aquilinum* and cross-leaved heath *Erica tetralix*. These preferences have been established by analysis of their droppings. Rabbits pass practically all their food twice through the intestine. It is passed down first to the rectum in soft pellets which are there collected by the rabbit and re-ingested. The hard pellets finally excreted were examined by C. R. Metcalfe during Lockley's experiment who found they consisted of various plant remains which he was able to identify. Among them were hairs of stinging nettles, showing that this plant is eaten but probably only by a hungry rabbit.

Rabbits draw foxes, stoats and weasels to heathland. When available rabbit flesh provides much of the food of these carnivorous mammals and all three will make use of a rabbit burrow for a den.

The fox *Vulpes vulpes* chooses a big burrow, enlarging it as necessary. From across the valley marshes we have watched fox cubs playing in the sun in front of such a den, with gorse for cover, the large opening to the run showing up dark against the yellow sand. One of the rare night sounds of the common has been the eerie wailing call of the vixen and the short answering bark of the dog. In pheasant-rearing seasons, a less pleasant reminder that there are still foxes on the common is the occasional carcase of a dead fox nailed to a tree.

A. L. Harrison Matthews (1952) writing before myxomatosis spread to Britain mentions an analysis of the stomachs of foxes which showed that during summer, rabbits were the commonest food (lambs coming second in sheep-rearing districts of Wales) and then small birds, insects, field-mice and voles. Shrews were not taken.

Stoats *Mustela erminea stabilis* and weasels *Mustela nivalis*, beside feeding on any small mammal or bird they can catch, are also partial to eggs, lizards and various insects. They, too, use smaller rabbit burrows for a home. On June 26th 1979, a family of weasels made their presence known on the Sandy Lane of Leiston Common when the arrival of a car interrupted the female moving her young from one part of the common to another, perhaps because her nest had been disturbed. She dropped the young she was carrying in her mouth in the middle of the lane not far from the bonnet of the car, and ran back to the undergrowth at the wayside where the last of her litter was waiting to be transported. Then she changed her mind and made a sudden dash back to the abandoned kitten and picking it up in her mouth by the loose skin of the neck, she trailed its long body across

WEASEL *MUSTELA NIVALIS*
BODY LENGTH EXCLUDING TAIL 165 mm

the lane and deposited it safely on the other side. Immediately she had
done so, she ran back and repeated the operation. The young weasels
were more sombrely grey than their small parent with her fine
chestnut brown back, bright white front and wide awake eyes and
ears. Female weasels are so much smaller than the male, averaging
half the weight, that locally they are thought of as another species and
called "mousers". (For the same reason, Harrison Matthews says
they are called "mouse weasels" in Scotland.) The young have a shrill
scream. They are weaned at four to five weeks and taught to hunt and
kill by their mother. (Maurice Burton 1976)

SOME BEETLE INHABITANTS OF OPEN HEATH

The grazed heath of the study area has an extensive underground
population—an immensely important section of the heathland
community. Around the entrances to rabbit burrows there are areas
of firm almost bare sand with perhaps nothing more than a scant
growth of sheep's sorrel *Rumex acetosella*. Here one often finds small
holes about the diameter of a little finger. In places these occur as
many as five to a square metre and are often covered with small
triangular mounds of loose earth.

Our curiosity aroused as to the inhabitants of these holes of which
there were so many in this patch of grazed heath, we adopted the

87

simple idea of sinking a jam jar near a rabbit burrow so that the rim was just below ground level. Rabbits do not soil their burrows but at a little distance the ground was covered with about sixty pellets to a square foot. It was not therefore surprising that the main catch was a beetle for whom rabbit or sheep pellets supply its every need—food and nesting material for itself and its young as well as a warm lining for its winter home.

MINOTAUR BEETLES

On November 7th 1978, trapped over-night in the jar was a large shiny black beetle, somewhat flattened in shape and hairy beneath with wing cases striated by horizontal raised lines. This was identified at the British Museum (Natural History) as a female minotaur beetle, *Typhaeus typhoeus*, one of the Scarabaeidae or dung beetles that was studied in detail by the famous French "insect man", Jean Henri Fabre.

It is the male that shows the outstanding characteristic of the three horns, one long horn each side of the thorax and one smaller one in the centre, pointing forward over the head. This "trident" Fabre suggests helps the male minotaur to crumble the rabbit or sheep pellets and cart his rubbish when constructing the nest. The female has only abortive knobs where the male has horns.

In late November the beetles are obviously still collecting food from the heath at night for by November 20th three males and three females had walked into the trap. These were transferred to a beetle enclosure, some 54 cms long and 29 cms wide, the glass sides being 4 cms apart and packed with moist sandy soil from the field dampened with rainwater. There was free access to a covered ledge at the top where pellets were spread for the beetles to collect. When touched or surprised, the beetles stay still and draw in their legs—the most vulnerable part of the body.

By the second night in the enclosure, the minotaurs, both males and females, had buried themselves and there was considerable earth shifting with earth piled up to the height of about five centimetres. From October onwards minotaur beetles each construct their winter home and parts of these shallow burrows were visible after the first ten days. One of these reached three-quarters down the cage, curving round at the bottom where a beetle was clearly visible. Then came a great change in temperature with heavy frost necessitating a thick blanket covering to the enclosure. (On the frozen ground of the heath

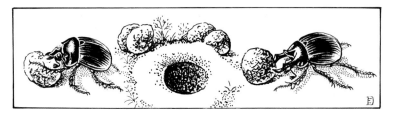

MINOTAUR BEETLES *TYPHAEUS TYPHOEUS*
BODY LENGTH 18 mm

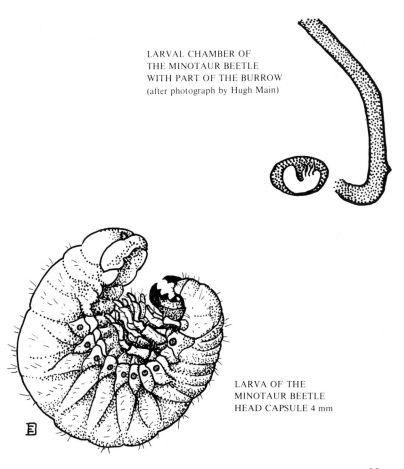

LARVAL CHAMBER OF
THE MINOTAUR BEETLE
WITH PART OF THE BURROW
(after photograph by Hugh Main)

LARVA OF THE
MINOTAUR BEETLE
HEAD CAPSULE 4 mm

outside, little conical heaps of debris covered the entrances to the burrows.)

During a few days of warmer weather from November to December 3rd, earth moving operations recommenced in the cage, showing the beetles had been out of the burrows. Pellets had been moved, in particular one pellet covered with the lichen *Hypogymnia physodes* had been shifted some distance and there was a general decrease in pellets on the surface. With the relatively mild weather continuing, obviously the minotaur beetles were about again on the heath as two more males were trapped in the jam jar on December 13th. They were fine large specimens with long horns and covered with pink mites. (These mites are probably co-feeders with the beetles eating any dung attached to the corselet and keeping the beetles scrupulously clean and shiny.)

From mid December 1978 there was some of the most severe wintry weather since the winters of 1947 and 1963. With some night temperatures registering −9°C and blizzards bringing shoulder high drifts of snow, the *Typhaeus typhoeus* experiment was interrupted and the ground trap removed until the end of frost and snow.

Since Fabre lost all his minotaur beetles in his cage during the severe winter of 1895, we were taking no chances and on the first day of March 1979 decided to introduce a new set of beetles into the enclosure. During the first two weeks of that month in a spell of warm spring weather, three males and three females were caught in two jam-jars placed near a rabbit burrow, at the foot of a mound of bell heather among moss and lichen with a thick sprinkling of pellets. The females were first introduced to the cage and allowed to settle. Then each male was dropped separately in one of the burrows the females had made, where they evidently settled for by the end of March increase in excavated earth showed there was considerable tunnelling and home-making activity.

At this period of the year, both male and female are engaged in about a month's intensive nest-making, the dung beetles being one of the few beetles where the male is equally involved in preparing for the young. The breeding burrows in the field are very deep—some going over eight feet (240 centimetres) down into the earth—so that there where it is cool and moist the shredded rabbit pellet never dries out. Fabre watched the mother, always ahead with her spade-like clypeus moving the earth and digging with her toothed arms. Her mate waits in the rear. His task is to take away the soil which he rakes towards him by the armful, kneading it into a plug which he pushes up and out of the burrow. Then the male sets about collecting and storing a little pile of rabbit pellets which when properly processed will provide food for the young in all stages of growth until they finally crawl to the surface as young adults the following autumn. Fabre describes how the male conveys the round pellet to the hole either by dragging it with its forefeet backwards or by rolling it with thrusts of the clypeus. Then clasping it with his legs, he inserts it in the hole. Here he lodges the pellet in the arc made by the three points of the trident, driving the fork in to keep it still while his forelegs with their toothed shanks are free to shred it into fragments. After this preliminary grinding, the fragments fall down the burrow to the female at the bottom who, writes Fabre, sets about "bread-making", patting the material with the flat forelegs, arranging it in layers and trampling it into the form of a sausage. The eggs will be laid just outside the larder about mid-April and take about a month to hatch into grubs. These grubs go through the whole summer feeding on the sausage of food until pupating.

The first two weeks of April 1979 was a period of intermittent sunshine with cold wind and some heavy rain. The tunnels now go right down the full length of the frame, showing a slightly oval dark mass of what must be mashed-up rabbit pellets at the curved bottom of the burrow. Is this Fabre's "sausage"? By April 13th, in the case of two burrows, the mass of provender about five inches long was now clearly visible. Collection of pellets remained constant at night, one entrance on one occasion being blocked with a pile of thirteen pellets.

Unhappily, during two days of absence from home (April 14th–16th) the beetle enclosure fell flat to the ground. (It had some days previously been re-arranged to stand in a bath of shallow water to prevent drying out.) A stop to all activity showed the male beetles at least had used this opportunity to get away. They would in any case have come to the surface to die when all preparations for the breeding burrow were complete. It is just possible that the females who live longer are still below. If they have laid eggs (Fabre's beetles did so around April 17th), since by now their store of food is more or less complete, young could be reared.

In fact no young minotaur beetles came to the surface in the Autumn but careful excavation of the soil in the enclosure revealed a larva, some 20mm long, lying on its back in a horse-shoe position in a small chamber at the bottom corner of the cage, coinciding with Fabre's finding that the egg is laid in the sand just outside the prepared victuals.

I have since learnt from R. D. Pope of the British Museum (Natural History) who kindly identified the larva, that the only recorded case of *Typhaeus typhoeus* being reared in captivity in Britain was by Hugh Main (1917) somewhat surprisingly using horse dung. He writes:

> The egg is deposited in the sand about half an inch from the first layer of food . . . The young larvae were first seen about the end of February and they eat their way to the top of their food mass and then back to their starting point when they are full-fed which was only reached in August 1917. Each then formed an oval pupation chamber in the sand beyond the old food mass, and then lay on its back to await pupation.

This, it seems, was the stage reached in the Leiston Common culture. Had development continued, the time taken from egg through larval and pupal stage to adult beetle would normally be five months.

OTHER BEETLES TRAPPED

Two ground beetles trapped with *Typhaeus typhoeus* were members of the family Carabidae—carnivorous hunters, swift in action and usually feeding at night on small creatures living in soil and moss. Alive in the jar where it spent part of the night of December 2nd 1978 was an entirely black beetle, identified by Howard Mendel (1978) of Ipswich Museum as *Nebria salina* which has been under-recorded in Suffolk perhaps because it has been confused with the more widespread *Nebria brevicollis*. The other beetle, several of which crawled into the jar, was the small metallic bronze beetle, *Notiophilus substriatus* about 4mm in length, greenish black underneath with bronzed black legs on which it can move very swiftly. Both larva and beetle are most lively in brilliant sunshine feeding on small creatures in the heathland soil. Imms (1947) says this beetle is rather local though widely distributed. Pill beetles *Byrrhus fasciatus* were also among those trapped. These vegetarians live in sandy soil at the roots of plants where their protective resemblance to seeds or "pills" of earth when they pack their limbs into a ball if disturbed, makes them hard to detect.

The heather beetle *Lochmaea suturalis* (see plate section) was around on Leiston Common and other Sandling heaths in the summer of 1979 and on summer evenings the green lights of glow-worms *Lampyris noctiluca* were seen on the verges of Sandy Lane. Eggs, pupa and males of the species all glow slightly but it is the females that shine brightly to attract males. The larvae feed by sucking liquid from paralysed slugs and small snails. The adults possibly do not feed at all.

At the end of February 1980, hot sunlight illumined branches of the pines on the fringe of the heath and showed the resinous pine cones to be thickly covered with large red ladybirds with seven black spots on the elytra. More typically associated with pine heaths in Suffolk are the eyed ladybird *Anatis ocellata*, large with yellow-ringed black spots, and a brown ladybird with white spots *Calvia quatuordecim-guttata*. Their larvae feed on scale insects and aphids.

The oil beetle *Meloë proscarabaeus* is the commonest of nine British beetles of the family Meloidae and being a beetle of grassy places and dry, warm climates, is often found on sandy coasts and heaths. It is a beetle whose strange life history is linked with some solitary bees that make nesting burrows in light soils where they sometimes act as host to the oil beetle larva. In its first active stage, this larva is a bright yellow louse-like creature little more than 1–2mms in length. According to Fabre (1882) on emerging from the egg it climbs up low-growing plants and wanders about in the flowers, waiting for any hairy insect that happens to visit them. The tiny larva then grips fast to the hairy visitor with its jaws and is carried away. If by any lucky chance it proves to be the right host and the bee carries it to its burrow in the soil, then the "triungulin" larva as it is called, first feeds on the bee's egg then on the pollen loaf the bee had prepared for its own young, meanwhile becoming less and less active as it passes through seven transformations before finally crawling to the surface as an adult beetle in spring.

On April 13th 1980, a male and female oil beetle *Meloë proscara-baeus* were found on Red Crag sand at Bawdsey Cliff in south-east Suffolk near where a variety of unidentified solitary bees were visiting their burrows in the sand. The female (32mm) was seen to be considerably larger than the male (20mm) even before becoming later swollen with eggs. Both were blue-black shot through with metallic purple tints, the thorax coarsely punctate. The antennae of both beetles are typically thickened in the middle and in the case of the male, the sixth and seventh joints are compressed giving a kink to the antennae. Both beetles had the abnormal elytra structure of oil beetles as described by Fabre: "Uncouth beetles with their clumsy belly and their limp wing cases yawning over their back like the tails of a fat man's coat that is too tight for the wearer."

Since these beetles are active in sunlight, they were placed by a sunlit window in a roomy dry glass aquarium draped with net to supply shade as needed and sprinkled regularly but sparingly with water. Turfs were planted in some 6 cms of sandy soil so as to leave areas of the soil bare.

The food of these oil beetles is entirely vegetable and they came immediately to dandelions, eating three large blooms in full pollen in one day. Buttercups were also eagerly taken, as were blades of growing grass. The female's appetite was at its height just before ovipositing.

NEST MAKING AND EGG LAYING

Attempted mating began almost immediately with the smaller male being carried long journeys on the back of the female from which position he continuously caressed her head with palps and antennae. Successful mating took place on April 22, marked by both remaining joined for over an hour, when the male having slipped off his mount was towed behind as the female clambered non-stop over the rough ground. This was followed by daily attempts by the male at mating, the female now much swollen with eggs appearing double the bulk of the small male. By April 27 two of the male's legs appeared useless and after slowing up generally, it died at the end of April. (Even in the wild the male has a very short life.)

After the death of the male, the female continued in restless movement except when voraciously feeding on buttercups and dandelions. On May 2nd, a hole was visible in the right hand corner of the cage near the sunlit glass, a good centimetre in diameter, going down perpendicularly and surrounded by a smooth turret of earth, distance from the bottom of the burrow to the top of the turret some 7 cms. Out of this, *Meloë* came to the surface tail first then managed to twist over without damaging the turret so as to disappear this time backwards down the burrow, the glint on the top of her head remaining just visible. When seen again, five hours later the burrow had been filled and the beetle had surfaced.

Preparations for a second batch of eggs took place on May 10th when the process was repeated but this time five abortive burrows were excavated before the final choice which was the only one to be filled with soil.

On May 22nd, it was noticed that the first batch of bright yellow eggs was visible through the glass, looking rather like a lump of hard boiled egg, some two centimetres across and the top lying about a centimetre below the soil surface. Through a magnifying glass it was possible to see that the cluster in fact consisted of hundreds of closely packed cigar shaped eggs about 1mm long, some of which seemed to be detaching themselves from the general mass with the embryo and black eye spots clearly visible inside. Thousands of eggs are laid by female oil beetles to compensate for the very low chance of survival of the young larva by finding the right host.

On May 23rd a third batch of eggs was laid and this time it was possible to see in detail both the excavation and the closing of the nesting burrow. Three preliminary holes were dug but abandoned (perhaps because there was insufficient depth of soil). Then the female started excavating on top of the first batch of eggs, digging down until she had scraped up a few eggs, when fortunately she stopped and started a new burrow alongside. Using her mandibles and clypeus, she rapidly scraped up the earth and pushed it under the thorax, collected it with the front legs and when a little pile had accumulated, brought it up to the surface moving up the burrow backwards until in a position to place the soil round the rim. This done, she dropped down the hole head first to continue the quite audible scraping down below. This process was repeated twenty times in forty-five minutes. When all was finished she came to the surface and as before reversed position to go down tail first in order to oviposit in the burrow now ready to receive the eggs. The metallic head could be seen below moving presumably with the exertions of egg laying then resting against the side of the burrow.

Meloë proscarabaeus remained in this position for nearly five hours. At 10.20 p.m. she raised her upper parts above the surface then with part of abdomen still in the hole and supported by two hind legs she pivoted round and round, methodically pushing the piled up soil down again into the burrow, using mandibles like a shovel, until the circular turret was flattened and the hole filled.

In about an hour the work was completed. The female's abdomen had contracted greatly, now no longer swollen with eggs, suggesting that a third batch had been successfully deposited. There are normally two batches but as in this case, three and four may occur.

On the last two days of life, June 2nd and 3rd 1980, *Meloë proscarabaeus* started excavating again even though both back legs had become useless, but in the middle of this fourth attempt at oviposition, the oil beetle died at the end of a reasonably long life since in natural circumstances after depositing her eggs, the female usually dies.

On no occasion while being handled were the beetles seen to exude the oily fluid (blood) that they are said to release when threatened and from which they get their name.

SUCCESSFUL HATCHING OF THE TRIUNGULIN LARVAE

TRIUNGULIN LARVA OF
OIL BEETLE
MELOË PROSCARABAEUS
BODY LENGTH 1·5 mm

On the afternoon of June 7th 1980, dark patches in the first egg clump, laid May 2nd, were seen to be moving and soon hundreds of minute orange, immensely active larvae were swiftly making their way one after the other through a tunnel in the sand to the surface, immediately climbing the glass walls of the cage. Under the microscope the bright eyes to the fore of the somewhat flat head are clearly visible as well as the claw and two setae on each of the six legs, and the two trailers on the hind tip of the segmented abdomen. From this tail end a drop of sticky liquid is exuded to help contact on slippery surfaces as the long body extends and contracts on upward movement.

Hatching of the first batch of eggs probably continued through-out the next six weeks, the stream of larvae climbing always upwards to the topmost heads of flowers, mostly Compositae, protruding above the nest side of the cage, which was now positioned out of doors where visiting bees were plentiful. Although placed among other flowering plants, the triungulin larvae tended to remain in a colony on the flowerheads growing out of the cage but their vigour was considerably renewed whenever fresh flowers full of pollen were introduced and to these they transferred immediately, crowding particularly on field scabious *Knautia arvensis*. They hide among the florets at night or when the weather is overcast but in sunlight run around or raise themselves on end, waving around in search presumably of a likely host. The larvae clung tenaciously by the mouth to a fine paint-brush when some were transferred to a mat of thyme in flower where bumble bees *Bombus lucorum* were gathering

96

nectar. They climbed five or six at a time onto the back of a bumble bee and remained on the hairs even when the bee tried to dislodge them. They also climbed onto a crab spider that had climbed into the cage, remaining on its back until it disappeared.

About a hundred of these triungulin larvae were still visible on the flowerheads fifty-five days after the first batch of eggs began hatching but then disappeared after a downpour. These minute creatures are so different from the large adult oil beetle that they were once thought to be a different species. Exactly what plants and which solitary bees they favour is not clear from the somewhat inconclusive earlier accounts. The Entomological Department of the British Museum know of no records this century of the observed hatching of the triungulin larvae or culture of *Meloë proscarabaeus* in Britain. The older and somewhat confusing accounts of Fabre and Newport mostly concentrate on a continental oil beetle *Meloë cicatricosus* and at the beginning of this century Dr Auguste Cros in France claimed to have succeeded in rearing another oil beetle *Meloë cavensis* completely on the rather solid honey of a certain *Anthophora* bee, slightly moistened, proving perhaps that the bee's egg was not the essential first nourishment for the triungulin larva. In the case of the *Meloë proscarabaeus* larvae observed in Suffolk, a few of these attached themselves by the mouth each side of a rather dry streak of honey on the side of a test tube but were engulfed if it became at all liquid.

BURYING BEETLES AT WORK

The remarkable burying beetles of the family Silphidae bury their carrion food and perhaps incidentally contribute to the sanitation of the heath. The females of these wide ranging beetles lay their eggs above the food store so that the larvae on hatching have only to make their way down to an ample larder from which they are at first fed by their parents. These beetles have a very specialised sense of smell and strong flight, and, while not specific to heathland their presence on Leiston Common was proved in the following way.

On March 17th 1979, a dead long-tailed field mouse *Apodemus sylvaticus* was carried to the study area of open heath where it was laid on the surface of raw sand scratched up by rabbits, then covered by a small wooden sieve raised on a few pebbles. Four days later there was no change in position but there were signs of loose earth being worked about ten centimetres from the mouse's head.

A week later, April 9th at 2.30, the sieve was again lifted this time to reveal a partially buried field-mouse—both hind quarters covered in sand up to waist level, with half of the tail still visible but curled to one side as though the haunches were buried in a doubled up position. The earth could be seen moving on the surface about five centimetres to the west of the mouse carcase, first in one direction then in another. Having no other tool to hand, I put my finger underneath and lifted out *Necrophorus humator*, one of the largest and most common sexton beetles, some 20mm long, all black except for distinctive and very handsome yellow clubs to the antennae.

Since the beetle was kept for identification, further developments might not have been expected but by 4 p.m. there was, in fact, spectacular progress. The mouse was now buried to the neck, with one front paw showing and two ears just above ground. By 6.30 one ear only and a small circle of face was visible, the whole now surrounded by a few centimetres high circular earthwork of excavated sand. By twilight an hour later, the work was completed. *Apodemus sylvaticus* was neatly buried surrounded still by an amphitheatre of earth which by the next morning was levelled flat.

Linssen (1959) gives seven centimetres as the depth at which burying beetles bury their carrion. Apparently the beetles dig with their front legs under the corpse, pushing aside the loose earth with the others until they make a mortuary into which the corpse sinks. Then they strip off the fur with their mandibles and compress the carrion into a ball-like mass of food for their offspring.

A month later the same experiment was repeated in the same spot but this time no burying took place. *Necrophorus humator* is still about on the commons, however, as the following June one flew into a light, its strong wings extended. A colourful red and black burying beetle *Necrophorus investigator* also comes frequently to the light in the study area. A complete red band with no yellow hairs on the thorax and yellow tips to its antennae are its recognition marks.

BEES AND WASPS ON HEATHLAND

BUMBLEBEES

The jam-pots sunk to catch beetles sometimes threw up surprises. In the late cold spring of 1979 during the first two weeks of April when the willow catkins were in full flower, three splendid specimens of large queen bumblebees wandered into the jars. These all had a lemon yellow collar band, another band of the same yellow on the black abdomen and bright white tails. They were identified by D. V. Alford as three queens of *Bombus lucorum*, among the earliest bumblebees to appear in spring as soon as the sallow and willow catkins provide food. They are very partial to the flowers of the heather family though they also visit other plants including brambles, clover, gorse and cherry.

Dr Alford suggested they were investigating the jar when in search of a hole in the ground for a nesting site. Only these fertilised queens live through the winter. They hibernate in the soil beneath leaf litter and it seems incredible that these bees could have survived the hard ground frosts of 1978–79. However, as D. V. Alford (1975) explains in his immensely informative book on the subject, hibernating bumblebees in severe conditions become cold-hardened by producing glycerol which acts as an anti-freeze and lowers the temperature at which ice crystals will form in the body.

Between hibernating and forming nests, the queens feed actively on pollen and nectar. The change to nest searching comes when ovaries begin to develop and hormone production changes occur within their bodies. Then they begin to scour over tussocky ground examining holes and tunnels for possible nesting places such as the disused nests of small mammals like shrews, voles and field mice.

One of the *Bombus lucorum* queens was observed for a day or two before release in a glass cage lined with soil, moss and cut grass, with diluted honey provided and a jar containing goat willow *Salix caprea* with male catkins in full bloom. It lapped the honey readily and busily investigated various tunnels in the moss and sand, combing the hairs of head and back free from debris after each excursion. It spent the night and part of the day in the tunnels when not exploring the willow for pollen.

Pollen provides the vital protein and a sufficient supply would have been essential for brood-rearing. This particular bumblebee is a pollen storer. The newly-gathered pollen is stored in the nest, either in waxen cells made for the purpose or in old cocoons, and from these it is fed by mouth to the larvae by the nurse bees. Nectar, a watery solution of various sugars, is also collected.

Bombus lucorum is abundant on heathland and other areas in every county of England and Wales and very plentiful in the study area of Leiston Common. Another bumblebee, *Bombus jonellus* sometimes called the Heath Bumblebee, is more specifically attached to heathland. There are a few records of the bumblebee in East Anglia but it has not been seen locally though it is widely distributed throughout Britain. The queens nest on or above ground. They are smaller than *Bombus lucorum* and they also differ in having two yellow bands on the thorax. The foraging bees visit members of the heather family (Ericaceae) and other heathland plants like broom *Sarothamus scoparius* and bramble *Rubus fruticosus*.

A third heathland bumblebee, *Bombus lapponicus*, is rarely seen on southern heaths but spends much time on mountains and moors where bilberry *Vaccinium myrtillus* is plentiful.

SOLITARY BEES OF THE HEATH

Whereas bumblebees live like hive bees in a community, other wild bees each make their individual nest. A large proportion of these solitary bees visit heathland for food and make nests in the sandy soil as do many species of solitary wasps.

A Nature Conservancy Council pamphlet on the conservation of these important pollinators by G. Else and others (1978) rates highly among their needs a nesting site of a warm sunny aspect with open unstructured vegetation allowing easy access to soil, and plentiful food supplies near at hand. With its southern aspect and its patches of semi-bare sandy soil near heather and gorse, the study area of Leiston Common would seem to fit the bill exactly so it is not surprising that when the warmth brings out the heathland scents the air is full of small bees and wasps coming and going from their solitary burrows in the sun-baked earth. Some of them are highly specialised in their food requirements and in this way help to avoid competition for the heath's resources.

Of the solitary wasps, two species have been watched in the study area. At mid-day on a rare hot sunny day of June 1979, a sand-wasp with a long narrow waist and a bright orange-red band across its abdomen was dragging what appeared to be a dun-coloured caterpillar over twice its size up the hard sand of a south slope sheltered from east and north by the garden wall. On the bank between patches of completely bare soil grew small plants like sand spurrey *Spergularia rubra* that thrive on being trodden upon in the hot dry soil. Here the wasp flew just above ground level with legs trailing, dragging the inert caterpillar over the ground. Prematurely perhaps, an attempt was made to cover both wasp and prey with a plastic container for observation, but the sand-wasp sped out between the rim and the

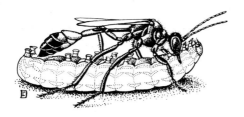

FEMALE RED-BANDED SAND WASP
AMMOPHILA SABULOSA
(BODY LENGTH 21 mm)
DRAGGING A PARALYSED CATERPILLAR

ground, leaving the caterpillar behind. Intent upon following the sand-wasp, its pursuers lost sight of the caterpillar which proved hard to find, its colour merging with the sandy earth.

During the ensuing hour, the wasp, a red-banded sand-wasp *Ammophila sabulosa*, returned several times in search of its prey which, perhaps because of human presence, it too failed to find. Early the same evening, however, a caterpillar was found lying paralysed but not dead on the ground about two inches from an open hole below it on the slope. On its side, because of its brown hairs, this caterpillar also looked dun-coloured but closer inspection showed its head and back were slate blue streaked with a bluish-white line. It was identified by H. E. Chipperfield as the larva of the lackey moth *Malacosoma neustria*. This moth lives in a group for most of its larval life and a nearby hawthorn bush, one of its food plants, probably furnished a fair supply. However the naked larva of Noctuidae is the more usual food of the red-banded sand-wasp as shown in the illustration.

Edward Step (1946) describes this wasp exactly as we saw it, straddling along dragging under its body a caterpillar bigger than itself, having first partially paralysed it by stinging it in several nerve centres. Each caterpillar is dropped in a prepared burrow and remains alive to provide fresh meat for one of the wasp's larvae. When the larder is full and the egg laid, the opening is firmly closed.

Hundreds of small nest-holes in the sandy soil of this area of Leiston Common are made by the field-digger-wasps *Mellinus arvensis*. The digging of each nest is the work of an individual female with no assistance from the male. She excavates the soil by using the

101

mouth as pick and shovel to carry sand to the entrance, always moving backwards so that the tail appears first out of the hole. (This observation was made by Henry J. Boreham (1951) who watched the operation in a sandhill at Bury St Edmunds, Suffolk.)

A female field digger-wasp *Mellinus arvensis* was caught at the entrance to its burrow in a sandy mound near a rabbit hole in the study area. Viewed under a microscope she was of handsome appearance—shiny black with a yellow line each side of the face, yellow collar, two yellow spots on the black segments of her black and yellow abdomen, and black and yellow legs to match. She fills the cells leading off from her burrow each with four to five blow-flies, hover-flies etc which she part paralyses then, with the prey tucked upside down under her legs, flies back to the nest.

Other wasps that burrow in sandy soil specialise in different prey. The sand tailed-digger *Cerceris arenaria* for instance only stores weevils, while the black-banded spider wasp *Anoplius viaticus* scrapes in bare sand with long black spiny legs in search of spiders, including the fat-bodied wolf spider *Trochosa terricola* that occurs on Leiston Common. She paralyses the spider without killing it, drags it to a suitable site then excavates a nest hole, filling each cell with a spider on which she will lay one egg. When all is completed she covers the entrance with heather litter, rabbit pellets or whatever is to hand.

Rare in East Anglia and not recorded in the study area, though widespread in the south, is an interesting wasp known as the heath potter wasp *Eumenes coarctata*. It constructs a pitcher-shaped nest of clay attached to the stem of heather or other woody plant. The pot consists of a single cell which is provisioned with sufficient small caterpillars (Lepidoptera) to serve as food for one larva.

BEES AND OTHER POLLINATORS ON A HEATHER BUSH

The solitary bees, like the solitary wasps, differ from their social relatives in that there is no worker caste and each female works on her own to provide home and food for her young. However, unlike the wasps the bees are not predatory and the larvae are not carnivorous. The young feed entirely on honey and pollen—a mixture known as pollen loaf. The need to collect this food makes these solitary wild bees of great value as pollinators particularly since some of them appear early in the spring before other species are about.

One afternoon, on Leiston Common in August of 1979, a single tall clump of mature heather *Calluna vulgaris* growing among a carpet of bell heather *Erica cinerea* was watched for visiting bees. Working the flowers in large numbers were small bees with head and thorax covered thickly with brown hairs, some paler than others. One of the latter was identified as a male *Colletes succinctus*, sometimes called by the English translation—"girdled colletes". It is confined to heather *Calluna* and *Erica* and can be seen swarming on heathland in late summer searching the flowers for nectar with its short divided tongue, the retaining hairs on the hind legs of the females bulging with pollen.

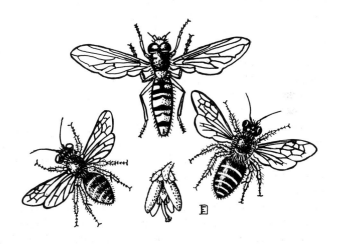

A HOVER-FLY, *EPISYRPHUS BALTEATUS* BL 11·5 mm
MINING BEES. *ANDRENA FUSCIPES*, MALE BL 9 mm
COLLETES SUCCINCTUS, MALE BL 10 mm

The female makes its nest in sandy soil in a straight shaft to a depth of some 25cms. This burrow is water-proofed with a thin membranous pellicle resembling cellophane produced from certain glands, then the cells are made one above the other, each filled with pollen moistened with honey on which is laid one egg.

Although the female *Colletes succinctus* intends her burrow solely for her own offspring, there is a little inquiline bee *Epeolus cruciger* that thwarts her purpose by watching an opportunity to slip inside the burrow and lay her egg in the loaf provided. This interloper's grub destroys the host's egg and then grows fat on the food in the larder.

The other solitary bees swarming on the clump of heather in

considerable numbers were small mining bees *Andrena fuscipes*, which visit heather flowers for nectar and pollen. Certain composite flowers like ragwort *Senecio jacobaea* also help sometimes to provide this provender which the females use to stock the many branched burrows they dig in the bare soil nearby. In this case also there is an unwanted intruder, a long waisted homeless bee *Nomada rufipes*.

On the particular afternoon in question, other insects visiting the heather bush included honey bees *Apis mellifera* workers, male and worker bumblebees *Bombus lucorum* and *B. pascuorum*, and a species of ichneumon wasp. Also sipping the heather nectar in the August sunshine were black and yellow hover-flies of wasp-like appearance. It was possible to identify three species: *Episyrphus balteatus*, a male and female of *Metasyrphus corollae*, and a male and female *Melanostoma mellinum*.

Heather is obviously an important food plant for a variety of insects but the benefits are by no means one sided. A Danish biologist O. Hagerup (1950) studied the importance of bees and other insects in the pollination of both heather or ling, and bell heather. Wind pollination is very important for both plants as one can see from the fine clouds of powdery pollen released from the masses of closely packed flowers but it would be impossible in many climates to rely on wind alone.

The structural features of the flowers of *Calluna* present a problem to some larger insects and Hagerup noticed that remarkably little pollen was seen on these insects that visited heather alone. This he thought was because when the heather flower is open the stigma protrudes so far out that the length of the style makes it difficult for larger insects to put their heads into the flower and reach the anthers which are confined to a very small space in the bell-shaped corolla.

However, there are minute insects about 1mm in length that provide extremely effective pollination. These are the heather thrips *Thaeniothrips ericae*, that perform all their life functions inside a heather bell—eating, copulating and egg laying. Being so small they can easily crawl into the narrow tube formed by the stamens where most of the nectary is hidden. To do this they cannot avoid touching the anthers and getting covered with pollen which gets shaken off onto the sticky stigma, particularly when the females struggle to open closely locked wings before flying off in search of a male.

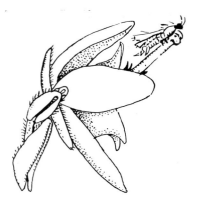

THRIPS *TAENIOTHRIPS ERICAE*
POLLINATING A HEATHER
CALLUNA FLOWER. ×12.

LONGITUDINAL SECTION OF THE FLOWER
SHOWING NECTARY (N). THE THRIPS
FORCES ITS WAY DOWN TO THE NECTARY
AT T1 AND T2 WHILE THE PROBOSCIS OF
A BEE IS INSERTED AT B ON THE LOWER
SIDE OF THE FLOWER WHERE THERE IS
GREATER SPACE BETWEEN THE COROLLA
AND THE STAMENS. ×15.

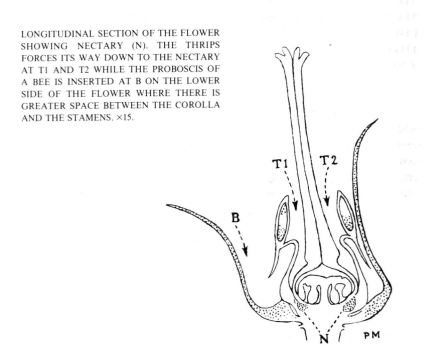

A SHORT LIST OF SOLITARY BEES ON HEATHLAND	PLANTS VISITED FOR NECTAR/POLLEN FOR LARVAE
Colletes succinctus (Colletidae)	Heather bloom *Calluna* and *Erica*
Epeolus cruciger (Anthophoridae)	Inquiline of *Colletes succinctus*
Andrena barbilabris (Andrenidae)	Mainly yellow Compositae flowers
Sphecodes pellucidus (Halictidae)	Inquiline of *Andrena barbilabris*
Andrena fuscipes (Andrenidae)	Swarms over heather *Calluna* and *Erica* which it visits for nectar and pollen. Occasionally visits yellow Compositae flowers
Nomada rufipes (Anthophoridae)	Inquiline of *Andrena fuscipes*
Andrena argentata (Andrenidae)	Visits *Calluna* and *Erica* blooms
Panurgus banksianus (Andrenidae)	Visits yellow Compositae
Panurgus calcaratus (Andrenidae)	Visits yellow Compositae
Lasioglossum prasinum (Halictidae)	Visits *Erica* and *Calluna*
Dasypoda altercator (=*hirtipes*) (Melittidae)	Sometimes swarms locally on heaths. Mainly visits yellow Compositae flowers
Heliophilus bimaculata (Anthophoridae)	Swarms locally on heaths, forming large, compact colonies. Visits heather bloom *Calluna* and *Erica*

SOLITARY WASPS ON HEATHLAND	PREY ON WHICH LARVAE FEED
Astata boops (Sphecidae)	Mostly immature bugs i.e. Pentatomidae
Psen equestris (Sphecidae)	Plant bugs—small Cicadoidea
Omalus panzeri (Chrysididae)	Inquiline of *Psen* species
Tachysphex pompiliformis (Sphecidae)	Immature grasshoppers (Acrididae)
Ammophila sabulosa (Sphecidae)	Caterpillars of moths including the larger, naked larvae of Noctuidae
Crabro cribrarius (Sphecidae)	Medium-size flies (Diptera)
Crabro peltarius (Sphecidae)	Diptera
Crossocerus wesmaeli (Sphecidae)	Small Diptera
Oxybelus uniglumis (Sphecidae)	Diptera
Diodontus minutus (Sphecidae)	Winged aphids
Mellinus arvensis (Sphecidae)	Diptera
Cerceris rybyensis (Sphecidae)	Small solitary bees i.e. *Halictus, Andrena, Lasioglossum*
Cerceris arenaria (Sphecidae)	Certain weevils i.e. *Otiorrhynchus* species
Methocha ichneumonides (Tiphiidae)	Tiger beetle larvae, *Cicindela* species
Myrmosa atra (Tiphiidae)	Various small wasps and perhaps bees
Pompilus cinereus (=*plumbeus*) (Pompilidae)	Spiders
Anoplius viaticus (=*fuscus*) (Pompilidae)	Spiders

Note: All the above bees and wasps lay their eggs in cells at the end of burrows dug in sandy heathland soils. The inquilines (Latin *inquilinus*=lodger) do not dig their own burrows but lay their eggs in the cells of other species.

GRASSHOPPERS OF THE STUDY AREA

Walking in early July over the springy turf between the springy clumps of young bell heather *Erica cinerea*, one's every footfall sends up spurts of minute leaping creatures—frog-hoppers in their myriads. Numerous, too, on sunny days are the leaping grasshoppers, at this time of the year still at various stages of development. With patience, it is possible to slip a glass tube over them as they momentarily come to rest and then one can see that they are mostly the mottled grasshopper *Myrmeleotettix maculatus*, the smallest of the common British species and typical grasshopper of drier heaths and moors. In early July of 1979 there were so many on this acre of open Sandlings heath that it was soon possible to catch three nymphs and one fully developed adult. The latter's fawn wings fully covered its abdomen but the growth of the immature wings in the three nymphs was at various stages corresponding to the number of moults they had undergone.

Under the microscope, the elegant herringbone pattern could be clearly seen on the swollen hind femur of the legs marking the lines of attachment of their strong jumping muscles. The colour combinations are astonishingly beautiful and varied. Even more astonishing was the way, in the four samples caught, the colouring of each resembled its own particular background, even though all were found at no great distance from each other. The nymph caught among the new sprouts of the rabbit-nibbled fringe of an old clump of ling *Calluna vulgaris* echoed exactly its vivid green and purplish red, with the black spotted pattern of the abdomen making it hard to see among the black twigs and peaty soil. Another found in an area of grey lichen and moss was whitish green and buff while the adult caught in short turf had head and pronotum mottled grass green and straw colour!

Certainly one would need to catch many more of these grasshoppers to ascertain whether this experience was general and not just coincidence but a leading authority on grasshoppers, D. R. Ragge (1965) found by experiment that the nymphs of the mottled grasshopper were able to alter their black pigment in a way that made them less conspicuous. However, in *Grasshoppers, Crickets and Cockroaches of the British Isles* he states that modern opinion tends toward the view that in most cases British grasshopper colouration is governed by genetic factors and not influenced by environment. Though much more work needs to be done on the subject, it is thought that natural selection explains why the colour varieties of the mottled grasshopper that harmonises best with a special background, are often most numerous in that particular place. It is these varieties that over the years on open heathland have escaped the eye of bird and lizard by their efficient camouflage.

Since both nymphs and adults feed on grass, it would be no good looking for them in dense heather where there is no open space with grass. These mottled grasshoppers need sparse vegetation of the open heath for breeding sites as well as for food. The female uses its ovipositor to dig a hole in the soil often choosing an ant's nest in which to lay its batch of eggs enclosed in a strong case or egg-pod. Six of our British grasshoppers lay their eggs just below the surface of the ground but of these only the mottled grasshopper *Myrmeleotettix maculatus* and the common field grasshopper *Chorthippus brunneus* would be likely to choose this very dry and acid bit of East Suffolk coastal heath.

SPIDERS ON HEATHLAND

On a dewy September morning when the vegetation is festooned at every level with sparkling webs, one gets some idea of the immense number of spiders that live on heathland. A Furzebrook Research Station report gives the number of different kinds of spider found on the heaths of south-east Dorset as two hundred—about a third of the total British species. The individual spider population of an undisturbed grass field near Bexhill has been calculated as not far short of two and a quarter million (Bristowe 1939).

Too much competition for the heath's resources is avoided in several ways. Different species have different preferences for a particular type of heathland whether bog, wet heath, or pine scrub. Others mature at different seasons and many species are adapted to feed on a special type of prey.

Typical of the open sandy heaths are the hunting and jumping spiders that instead of waiting for insects to get entangled in snares, go out in search of prey, relying on their power to leap and run. A jam-jar trap sunk at the mouth of a rabbit-hole near bell heather, lichen and moss on the dry heath of the study area, collected seven different kinds of heathland spiders in March and April of the cold spring of 1979. One of these was a light brown spider, a female

Agroeca proxima of the family Clubionidae, that lives under heather or in a clump of grass. Bristowe (1958) describes how she encloses her eggs both in a cone-shaped inner covering of silk, and a bell-shaped outer covering. The white silken bell is suspended from a heather twig and is later covered with a thick layer of earth. These spiders are mostly active at night for despite the two curved rows of eyes, these short-sighted night hunters rely largely on a sense of touch to stalk their prey.

Peter Merrett and his team, in experiments over three years on selected areas of Dorset heaths, found *Agroeca proxima* was among the most numerous and widespread of heathland spiders, being most active from August to September with a peak of activity in October.

Also taken near the rabbit hole on Leiston Common were two male wolf spiders—*Trochosa terricola*—a species that is very common on dry heaths where they live among heather during the day and prowl at night looking for insects. Another mainly nocturnal hunter trapped was a female *Zelotes latreillei*, a shiny black spider covered with short velvety hairs.

A female wolf spider (*Pardosa* sp.) was found in the study area at the base of a low growing clump of bell heather carrying its egg case, as is usual in the species, on its spinnerets. It was transported to the house of the artist to be drawn on August 5th 1979. Three days later, there was a surprising transformation in appearance caused by the fact that the

FEMALE WOLF SPIDER *PARDOSA* SPECIES (BODY LENGTH 5 mm) WITH SPIDERLINGS AND EGG SAC

young had hatched (presumably through the holes seen in the "seam" of the egg sac) and had clambered on to the parent's back. There was no sign of movement in the spiderlings and the egg sac remained attached, deflating gradually during the morning. At mid-day the spider could be seen detaching the sac with its hind legs. This sac the artist described as of a light warm red-brown with paler seam.

By the next day the three young spiders had left their parent on short expeditions but all returned by nightfall or immediately when frightened, using their parent's back leg as a ramp. The female transports its offspring in this manner for about the first week of life.

All the young spiders that came into the jam jar were species that are active in bright sunshine: three juvenile wolf spiders *Pardosa palustris* that hunt at great speed in full sunlight and some jumping spiders of the genus *Salticus* that are attracted to hot sunny places. When the sun is not shining they remain in silk cells under plants. These jumping spiders stalk and jump on their prey like a cat. They walk in jerks and can swivel their heads round so that, with their battery of eyes, they can watch their prey from every angle.

One of several very small spiders trapped was a slim striped little spider *Bathyphantes gracilis*, belonging to the immense family of money spiders Linyphiidae that hang in autumn and winter beneath their hammocks low down in the heather and grass until one day they feel the urge to climb upwards and float away to where the air is cooler. Squeezing silk from their spinnerets they wait for upwards air currents to waft them on gossamer threads up and into the distance. Bristowe has an excellent detailed description in the *World of Spiders* (1958).

Pitfall trap records over three years showed the following spiders were among the most frequent on dry heaths in Dorset (P. Merrett 1967). All are hunting spiders with the exception of the first:

Atypus affinis (Atypidae) *Xysticus sabulosus* (Thomisidae)
Dysdera erythrina (Dysderidae) *Xysticus cristatus* (Thomisidae)
Agroeca proxima (Clubionidae) *Attulus saltator* (Salticidae)
Clubiona trivialis (Clubionidae) *Euophrys petrensis* (Salticidae)
Scotina gracilipes (Clubionidae) *Oxyptila atomaria* (Salticidae)

A GORSE-BASED COMMUNITY

At one corner of the open heath study area, the complex three dimensional structure of a long low gorse bush (*Ulex europaeus*) provides a specially safe living place for a variety of creatures. Rabbits have created a small warren beneath it and their constant nibbling has sculptured the bush into a sort of Henry Moore reclining figure some eight feet long, rising to three feet high, with knobs and bulges of dense prickly spines. On its "back" a scattering of pellets at all levels show that the rabbits climb its full height. At all levels too the spiders are at work, making maximum use of every possibility by slinging hammocks of mesh webs and vertical orb webs in every nook and cranny. Some stretch their scaffolding only from the spines at the ends of branches. One little comb-footed spider *Theridion sisyphium*, constructs scaffolding snares in gorse where the female weaves a small silk cone or breeding tent under which she presses her bundle of eggs and here she remains until the eggs are hatched. She is about 4mm in size, her round abdomen neatly patterned with white on shining black.

When on July 15th 1979 a branch of gorse *Ulex gallii* complete with this spider and her web was put in a container for observation, perhaps because its position had been altered less to her liking, *Theridion sisyphium* removed her bundle from the tent and attached it to the dark lid of the container. She then hung upside down under the egg parcel, partially steadying it with her middle legs. The round bundle was as big as herself and covered with special greenish silk. When the container lid was removed, despite the alarm and disturbance, she remained continuously with the eggs, only moving to dart after the package when it became dislodged. G. H. Locket in 1926 first witnessed this little spider feeding her young from her mouth while hanging upside down below the breeding tent, behaviour found only in a few spider species.

In the same month of July another member of the Theridiidae family was seen tackling an eleven spotted ladybird *Coccinella 11-punctata* on the rabbit-nibbled gorse bush. She moved round this difficult customer presumably letting out silk from her spinnerets. During this time, she repeatedly turned her back on the struggling ladybird and appeared at first to be kicking it. In fact she was using her hind legs to throw silk over her prey and bind it.

It was the gale of August 14th 1979 that wrecked the Fastnet yachting race and lashed the whole common here into frenzied movement that sent me to the same gorse bush to see what had happened to the elaborate structure of a spider I had been watching for the previous two weeks. With every scrap of vegetation tearing at its roots before a fifty miles-per-hour gale, how had it survived?

The structure had consisted of a long funnel of spun silk widening out to a dinnerplate sized platform stretched between two branches with a maze of overhead scaffolding threads. The funnel led deep into the prickly safety of the gorse and into its depth, whenever I arrived, sped at a great pace a large spider whose abdomen I nevertheless distinctly saw on three occasions. Its markings, a pale inverted fishbone pattern running the length of the top with darker vertical markings each side, showed it was almost certainly a female *Agelena labyrinthica*.

I found most of the previous structure had been demolished. What remained appeared to be a tangled ball of web blocking the entrance where the funnel had been, with no visible sign of the spider. The following day there was still no sign of life but by the second day she had clearly been forming two or three little passages in the tangle and, on the third day, one opening had been enlarged enough to see *Agelena labyrinthica* moving round and round what, by putting one's head as far as bearable into the gorse, could quite clearly be seen to be a large cream-coloured oval "box" about three-quarters to a half inch in size and mottled with brown markings. This was hanging above a lower chamber and already attached to the sides of the passage. The spider, this time taking no notice of the face peering so near, was crawling round and round the egg chamber, spinning silk from her spinnerets and every now and again pausing to dab her "tail" against the sides of the passage to secure it. This activity continued for most of the afternoon.

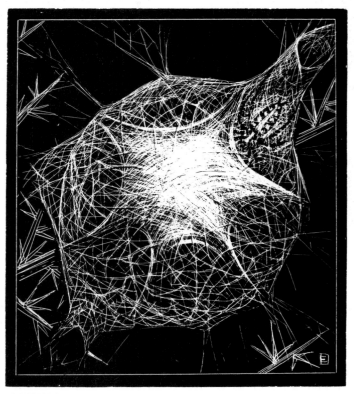

NURSERY OF *AGELENA LABYRINTHICA*

The next day the job was obviously finished. She had constructed round her egg-case the impressive breeding chamber whose labyrinthine passages gave her her name. Five days later she had reconstructed at its entrance a much smaller platform, with a maze of trip wires above it against which greenfly and one larger true fly had already been caught and fallen on the sheet below. One could expect the eggs to hatch sometime in October.

It appears that a whole number of spiders are found on gorse bushes more often than elsewhere, and the same ones are found whether the gorse is growing in heather country or grassland, although, of course, there are distinct regional differences.

A MOISTER HABITAT

Many other creatures besides spiders make their home in gorse. In June for instance the bush described is full of flitting micro-moths, two of which were identified by H. E. Chipperfield as a male light striped-edge piercer *Laspeyresia succedana* and a female dark crescent piercer *Grapholita internana*. Their caterpillars live in the gorse pods and feed on the seeds. The larvae of two small weevils, the pear-shaped *Apion striatum* and *Apion ulicis*, the gorse weevil, also spend

GORSE WEEVILS *APION ULICIS*
FEMALE BODY LENGTH 3·25 mm, MALE 2·5 mm
(BOTH SEXES VARIABLE IN SIZE)

their whole immature existence in the pods. The female weevil bores a hole with her snout into the young gorse pod then lays about half a dozen eggs inside it. The larvae feed on the seeds until they turn into pupae. When the adult beetles emerge they wait until the sun ripens the pods fully, then on one of those dry hot days when the whole common is full of little popping explosions, the pods burst open and the weevils are flung into the air in the place of the seeds they have eaten. Another insect, the green and brown gorse shield bug *Piezodorus lituratus* in years when it is plentiful remains in full view on the furze needles sucking from the branches the plant juices.

113

The gorse bush provides not only evergreen shelter and a prickly safe refuge but shade and moisture that allows creatures to live on the heath that might not thrive in the drier, barer areas. Small mammals seek its shelter for some of these reasons. By an unfortunate accident one January night we discovered that a pigmy shrew *Sorex minutus* that made its home beneath the bush, had fallen into a jar sunk on the edge of the gorse to see what beetles travelled there. The record describes it as a pigmy shrew with body 35mm, tail longer than the body at 40mm, hind foot 10mm and a long sensitive nose with wide nostrils. The compact velvety fur brushes either way like a mole's to whom this little insect eater is related. Ipswich Museum confirmed from its skull that it was indeed *Sorex minutus*. When J. C. Pernetta (1976) of the Animal Ecology Research Group, Oxford, analysed the guts and stomach of a pigmy shrew on Rough Common, Wytham Hill near Oxford, he found all spiders eaten were of the family Linyphiidae (money spider) of 2–3mm or smaller. These together with very small beetles and larvae were its main items of diet. Even the shock of finding itself trapped, let alone the cold, could kill a shrew so since *live* shrews were wanted on Leiston Common the jar was removed forthwith.

PIGMY SHREW *SOREX MINUTUS*
BODY LENGTH EXCLUDING TAIL 56 mm

Another small mammal utilises the double shelter of a short piece of corrugated iron sheeting that has lain under the gorse bush for many years. It left evidence of its existence in several runs and a small pile of the seeds and old flower-heads of bell heather *Erica cinerea*, together with many wing cases of dismembered minotaur beetles *Typhaeus typhoeus*. Pieces of cheese, apple and banana provided were taken on three separate nights pointing to a more omnivorous eater than the shrew. This turned out to be a male long-tailed field mouse, *Apodemus sylvaticus* with distinctive long ears, tail a little longer than the body, which was brown above and light silver grey below with a small patch of yellow on the breast between the arms.

The banana pieces were found covered with ants identified by A. Abbott of Furzebrook Research Station as *Lausius niger*, a black ant that is a common grass and heathland species preferring damper areas of heathland. It mainly forages on the surface for flies and plant bugs, and collects nectar both from heather flowers and from the nectaries of young bracken fronds. (Three such curled fronds were pushing their way up under the tin.)

One burrow here puzzled us for a long time—a sort of squared slit in the soil that gradually widened until one morning at the end of July there was a dark-coloured warty toad *Bufo bufo* with copper rimmed pupils to its eyes gazing out of the hole, head and curved legs alone visible. Here it sat snapping up insects, clearly finding this moist and safe tin shelter to its liking.

BUTTERFLIES AND MOTHS: CAMOUFLAGE FOR SURVIVAL

For creatures of the open country, the ability to merge into a background is sometimes essential for survival and heathland populations present many examples of such successful camouflage. Butterflies and moths, for instance, often surprise by their astonishing capacity suddenly to hide bright colour patterns and disappear into their background.

The small heath *Coenonympha pamphilus* is a very common and widespread butterfly of the Suffolk Sandlings. The female used to be known as golden heath eye (South 1973) and when the August brood emerges and dozens of these small butterflies are frolicking in sunshine over the heath one understands why. But once let them settle wings closed, to rest on a twig of heather or stem of dry grass, as they do towards evening, and they appear no more conspicuous than a small faded leaf. This is because only the slatey grey and brown of the undersides of the wings are now showing.

Such camouflage is especially marked in the case of the grayling *Hipparchia semele*, usually common on all east Suffolk coastal heaths including Leiston Common and the Crown Walks, where it visits heather flowers for nectar. This butterfly has a habit of spending much time sunning itself on the ground with closed wings so tilted as to show the whole underside of the hind wing and the tip only of the forewing, while casting the minimum shadow. Between brief flights it sits in this position sheltered from the wind on the gritty twig-strewn ground between clumps of heather, toning perfectly with the greys and brown of the peat. Its mottled and striped caterpillars feed on various heath grasses, including tufted hair-grass *Deschampsia caespitosa* and early hair-grass *Aira praecox*.

In the case of the green hair-streak butterfly *Callophrys rubi*, the delicate green underwings merge with the leaves on which it alights, wings closed. It favours boggy heath but it is found on most Suffolk heaths. It lays its eggs singly on the petals of furze *Ulex europaeus*, in the flower heads of bird's-foot trefoil *Lotus corniculatus*. Other food plants include broom *Sarothamnus scoparius*, the berries of buckthorn *Rhamnus* and the flower-heads of mouse-ear hawkweed *Pilosella officinarum* and heather *Calluna*. Where these plants grow, one can hope to see this beautiful butterfly.

The small copper *Lycaena phlaeas*, with its fiery copper-coloured wings and lively behaviour, is attracted to acid heaths like Leiston Common where sorrel *Rumex* grows on which it lays its eggs to provide food for its young. Its caterpillar is more or less sorrel-green when fully grown, with pink markings that appear to mimic the red colouring often found on its food plant. When hibernating, according to South, it changes colour to dull olive and then when winter ends it once more changes back to the spring green of sorrel.

The silver-studded blue *Plebejus argus* is above all a butterfly of sandy heaths, requiring the warmth and direct sunshine they provide for successful mating. When fully grown the larva is reddish brown, with a black stripe on back and sides edged with white, making it quite difficult to find among the heather which is one of its food plants. While the female butterfly is sooty brown, the male is purplish blue with a black border on the outer margins and silver-studded blue spots on the underside of the hindwings. It is mainly of southern

AUTHOR'S HOME AND STUDY AREA, LEISTON COMMON

THE HOME OF THE GORSE-BASED COMMUNITY

THREE
EAST ANGLIAN
SHEPHERDS

HEAD SHEPHERD
BROWN, 1812–1892.
THE TOP HAT
SIGNIFIES
HEAD SHEPHERD

JOHN WEBB
SHEPHERD AT
THORINGTON HALL.
ESSEX, 1921

ARTHUR SUTTON,
SHEPHERD ON
THE SUDBOURNE
HALL ESTATE, SUFFOLK.
1919–1969

WESTLETON BEDS, SUFFOLK

WESTLETON WALKS, SUFFOLK

MOUSEHOLD HEATH, NORFOLK (Painting by John Sell Cotman)

PATTERNED LAND ON A HEATH NEAR WEETING, NORFOLK. UNDERLYING PERIGLACIAL "SOIL STRIPES" ARE REFLECTED IN THE VISIBLE PATTERN OF VEGETATION, DARK STRIPES BEING CAUSED BY THE GROWTH OF HEATHER *CALLUNA VULGARIS* ALONG TROUGHS OF ACIDIC SAND IN OTHERWISE CHALKY SOIL WHERE IT GIVES WAY TO THE LIGHTER COLOURED GRASS.

HEATHER BEETLE
LARVA

FULLY-GROWN HEATHER *CALLUNA VULGARIS.*
AN INCREASINGLY RARE HABITAT

HEATHER BEETLES ON A SANDY TRACK, DUNWICH HEATH, MAY 1980

FIELD OR SHORT-TAILED VOLE *MICROTUS AGRESTIS*, LEISTON COMMON

NIGHTJAR NESTLING ON A SUFFOLK COASTAL HEATH

distribution and, unlike the ubiquitous common blue *Polyommatus icarus*, has not been seen in the study area though populations appear to be re-establishing themselves, even if still sparingly, on Westleton and other heaths of the Sandlings.

The colour changes in the larvae of one of our most striking heathland moths seems to fit in with the seasonal pattern of heathland vegetation. The large and handsome emperor moth *Saturnia pavonia* has been recorded flying in daytime over several of the Suffolk coastal heaths though it has not been recorded on Leiston Common. Among its favourite food plants are heather *Calluna*, bramble *Rubus* and blackthorn *Prunus spinosa*. At an early stage the caterpillar is black, changing to black with orange rings, then when the greenery of the plants is at its brightest, it too is bright green with black markings so that it matches well the green leaves and black twigs of the heath.

THE COCOON (40mm) OF THE EMPEROR MOTH *SATURNIA PAVONIA* WHICH ON BEING DISSECTED REVEALED THE REMAINS OF THE CATERPILLAR CONSUMED BY THE LARVAE OF THE ICHNEUMON FLY *AGROTHEREUTES SATURNIAE* (BOIE), A FAIRLY COMMON PARASITE OF THE EMPEROR MOTH. THIS SPECIES IS PROBABLY COMPLETELY HOST-SPECIFIC ACCORDING TO M. R. SHAW OF THE ROYAL SCOTTISH MUSEUM WHO IDENTIFIED IT.

THE FEMALE ICHNEUMON STICKS ITS OVIPOSITOR THROUGH THE NEWLY MADE COCOON TO STING AND PARALYSE THE HOST LARVA INSIDE AND THEN DEPOSITS SEVERAL EGGS ON IT. WHEN THE PARASITE LARVAE HATCH THEY QUICKLY CONSUME THE HELPLESS HOST AND THEN MAKE THEIR OWN FRAIL COCOONS WITHIN THE TOUGH PROTECTIVE COCOON OF THE EMPEROR MOTH IN WHICH THEY OVERWINTER. THE ARTIST NOTICED THE ICHNEUMON FLIES APPEARING FROM JUNE 9th ONWARDS. (THE SMALL COCOONS FROM WHICH THEY HAD ISSUED CAN BE SEEN HERE JUST BELOW THE EMPEROR MOTH LARVA.) NORMALLY THE PARASITIC BROOD CHEWS A NUMBER OF SMALL ROUGHLY CIRCULAR EXIT HOLES IN THE HOST COCOON AND IT IS ALSO POSSIBLE THAT A FEW FORCE THEIR WAY OUT OF THE "VALVE" THAT THE MOTH WOULD HAVE USED.

The true lover's knot *Lycophotia porphyrea*, an elegant little moth with white markings on purplish red, frequents Leiston Common and most bell heather and ling clad heaths and commons. Its larvae feed only on these heather plants, hiding by day in the moss and litter below. This caterpillar, too, with its reddish brown colour, purplish stripes and black markings is well camouflaged in the heather and hard even for a bird to see.

MOTHS OF THE STUDY AREA

One delight of making a study of a special area of heathland has been to make the acquaintance of a great number of nocturnal inhabitants whose presence and whose beauty one had only previously guessed at. These are the night flying moths of Leiston Common and its environment, about one hundred and fifty species of which came to light between June 20th and November 30th 1979. The subtlety of colour, texture and design is beyond imagination. Once more the importance of the plant community in determining the make-up of the heath's animal population is underlined. Each of the main plants in the area attracts its own group of moths whether heather, bracken, bedstraw, bramble, lichen, ragwort, gorse, broom, birch, pine or grasses.

Below, some of the moths of Leiston Common are grouped according to which of these plants are used as one item of larval food. (Most of these moths have several food plants, however, and for further details see South (1977 edition). Many are not specific to heathland but range widely wherever their food plants are found.)

THE TRUE LOVER'S KNOT
LYCOPHOTIA PORPHYREA
WING SPAN 30 mm

118

HEATHER FAMILY *Ericaceae*: True Lover's Knot *Lycophotia porphyrea* de Vill., Fox Moth *Macrothylacia rubi* Linn., Mottled Beauty *Alcis repandata* Linn., Great Brocade *Eurois occulta* Linn., Narrow-winged Pug *Eupithecia nanata* Prout., Plain Wave *Idaea straminata* Borkh., Heath Rustic *Xestia agathina* Dup., Common Heath Moth *Ematurga atomaria* Linn., Autumnal Rustic *Paradiarsia glareosa* Esp., Beautiful Yellow Underwing *Anarta myrtilli* Linn.

BEDSTRAW *Galium* ssp: White-line Dart *Euxoa tritici* Linn., Silver-ground Carpet *Xanthorhoe montanata* D. and S., Square-spot Dart *Euxoa obelisca* D. and S., Barred Straw *Eulithis pyraliata* D. and S., Green Carpet *Colostygia pectinataria* Knoch., Archer's Dart *Agrotis vestigialis* Hufn., Cream Wave *Scopula floslactata* Haw., Flame *Axylia putris* Linn.

BRACKEN *Pteridium aquilinum*: Brown Silver Lines *Petrophora chlorosata* Scop., Angle Shades *Phlogophora meticulosa* Linn., Gold Swift *Hepialus hectus* Linn.

RAGWORT *Senecio jacobaea*: Cinnabar Moth *Tyria jacobaeae* Linn., Lime Speck Pug *Eupithecia centaureata* D. and S.

SORREL *Rumex* spp.: Blood Vein *Timandra griseata* Peters., Bird's Wing *Dypterygia scabriuscula* Linn.

BRAMBLE *Rubus* spp.: Lesser Broad-bordered Yellow Underwing *Noctua janthina* Sch. Dotted Clay *Xestia baja* D. and S., Brown-spot Pinion *Agrochola litura* Linn.

GORSE *Ulex europaeus*: Cream-spot Tiger *Arctia villica* Linn., and various micro-moths.

BIRCH *Betula*: Buff Ermine *Spilosoma luteum* Hufn., Grey Dagger *Acronicta psi* Linn., Double Square Spot *Xestia triangulum* Hufn., Pebble Hook-tip *Drepana falcataria* Linn., Early Thorn *Selenia dentaria* F., Feathered Thorn *Colotois pennaria* Linn., Large Emerald *Geometra papilionaria* Linn., Light Emerald *Campaea margaritata* Linn., November Moth *Epirrita dilutata* D. and S. Coxcomb Prominent *Ptilodon capucina* Linn., Broad-bordered Yellow Underwing *Noctua fimbriata* Schreber.

BROOM *Sarothamnus scoparius*: The Broom Moth *Ceramica pisi* Linn., The Streak *Chesias legatella* D. and S.

SCOTS PINE *Pinus*: Bordered White *Bupalis piniaria* Linn., Grey Pine Carpet *Thera obeliscata* Hubn.

WILLOW *Salix*: Pale Prominent *Pterostoma palpina* Clerck., Mouse *Amphipyra tragopoginis* Clerck., Sallow Kitten *Harpyia furcula* Brahm., Pink-barred Sallow *Xanthia togata* Strom., Least Yellow Underwing *Noctua interjecta* Hubn., Red Underwing *Catocala nupta* Linn.

GRASSES *Gramineae*: Dark Arches *Apamea monoglypha* Hufn., Straw Underwing *Thalpophila matura* Hufn., Smoky Wainscot *Mythimna impura* Hubn., Drinker *Philudoria potatoria* Linn., Common Wainscot *Mythimna pallens* Linn., Hedge Rustic *Tholera cespitis* D. and S., Rustic Shoulder-knot *Apamea sordens* Hufn., Common Rustic *Mesapamea secalis* Linn., Rosy Minor *Mesoligia literosa* Haw., Antler Moth *Cerapteryx graminis* Linn.

LICHEN and ALGAE: Rosy Footman *Miltochrista miniata* Forst., Common Footman *Eilema lurideola* Zinck.

MOSS *Musci*: Scarce Footman *Eilema complana* Linn.

As was expected moths coming to light in the study area included several that are mainly of coastal distribution like the bright-line brown eye *Lacanobia oleracea* Linn., the dark spinach *Pelurga comitata* Linn., a moth of waste places near sandy coasts, and the yellow belle *Aspitates ochrearia* Ross., that frequents rough coastal meadows and the Breck.,

119

ADDERS, SLOWWORMS, LIZARDS

The Indian summer of August—September 1979 was memorable for the opportunity to watch adders every day for nearly three weeks during a period when the small heath butterflies were swarming everywhere over the heath, visiting the bell heather flowers for nectar and chasing each other in spiralling flight. It was in mid-August while watching the butterflies that I found a cast skin of an adder *Vipera berus* some two feet in length lying on a mound of bell heather *Erica cinerea* far from public footpaths on the fringe of the open heath study area. Most adders migrate in Spring from their winter quarters on dry heath to wetter land but some stay throughout the summer so the find was not surprising. What did surprise me was the sight that followed later.

The heather came late into bloom after a cold spring and was still in full flower on August 20th when, again in search of insects visiting the bell heather, I saw on precisely the same mound near a patch of twisted dead heather wood, two adders facing each other, vertically coiled with heads and upper parts raised, looking astonishingly golden against the vivid purple heather and green heath grasses. As I paused in some wonder at the sight about a metre away, the smaller of the two with the darker zigzag markings uncoiled and slid with perfect economy of movement under the heather. The larger remained still for a full minute, and then followed suit. This larger adder was beautifully fresh in appearance, the zigzag markings of a rich nut brown on the greenish yellow body. (According to E. N. Arnold of the Natural History Museum, Department of Ecology, it was almost certainly a female.)

The following day at noon, the adders were in the same place, coiled this time flat against the ground, a complicated pattern of broken circles and zigzags making a perfect camouflage against the blackened twigs of heather. I approached them this time as before, (but now clad in wellington boots!) quietly from the north pausing about the same distance away. As before, the smaller, possibly a male, slipped immediately away below but this time the heavier female stayed lazily where she was in the hot sun, and was still there three minutes later when I left.

The next day in mid-morning the adders were in the same place, this time intertwined coil upon coil, in sleep. The female who was nearest to me merely showed she was aware of my presence by pushing her head further under her companion making it possible to compare their colour differences. After five minutes, there was no change in position and when I left they were both sleeping in full view in the sunlight, with white-tailed bumblebees working the heather around them. Two hours later when the sun had gone behind the clouds, the smaller darker adder was there alone and, by now very alert and awake, it slithered immediately under the heather.

Every day, except during rain of which there was very little, the adders were seen and photographed in the identical spot on the bell heather mound. Most days only the larger female remained long enough to photograph and during the last week she alone was seen. On one occasion she was sunning herself as late as 4 p.m. but she grew increasingly wary. After the first week in September, she disappeared entirely from the spot and was not seen on any succeeding day that month.

Adders *Vipera berus* are the most typical reptiles of dry heaths. In Britain they mate in April and May and the young are born in early September. Malcolm Smith (1951) says that pregnant females in July and August may bask all day, choosing the most exposed places in which to lie and returning to the same spot day after day for weeks at a time. I met the same female ten days later (September 16th) some twenty metres north of the usual site on the open heath. We looked at each other for some few seconds then she wound her way over crushed bracken to a bramble and gorse thicket under which she disappeared. Perhaps she had given birth and under the shelter of the bushes would use a rabbit hole in which to hibernate in company with others. I did not investigate further. Adders are not aggressive if left alone and not frightened, but *left alone* is the operative phrase for their bite is notoriously poisonous. They use this venom to kill their favourite food—common lizards, slowworms, frogs and baby mice.

Slowworms *Anguis fragilis* occur in the low undergrowth between the heath and garden wall where they glide with astonishing ease. These neat shiny harmless creatures with their very smooth scales are, of course, not snakes but a sort of legless lizard. They burrow in light soil and come up to the surface at dusk to forage for their favourite small white slugs, insects and larvae. Their young are born in a bag out of which they break by a stabbing movement of the head. A large slowworm kept by children on Leiston Common under conditions advised by the Natural History Museum, lived healthily for two years before being released.

The hot sandy soil of the common is well liked by lizards but only the viviparous lizard *Lacerta vivipara* has been seen, and that only occasionally when sunning itself on garden walls. The belly of the male is deep orange spotted with black, that of the female altogether paler. As the name implies, the young are born alive, the newly born being almost black in colour.

1 Grass Snake *Natrix natrix*: fully grown male 3ft=914mm., females larger. Habitat where ponds and ditches are plentiful in hedgerows, open woodland and some heaths near ponds.

{2 Adder or Viper *Vipera berus*: male.
{3 Young female adder.
A fully grown female adder is larger than average male which can attain 2ft=609mm. Frequents undergrowth on heaths. Zigzag not always clearly defined and ground colour variable in both male and female.

4 Smooth Snake *Coronella austriaca*: in England seldom exceeds 2ft=609mm. Prefers dry southern heaths and open woodland.

{5 Viviparous or common lizard *Lacerta vivipara*: male.
{6 Young common lizard at birth.
{7 Underside of *Lacerta vivipara*.
Grown male common lizard 6in=150mm., female often larger, with tail more than half of total. Heaths, banks, walls, etc.

8 Sand Lizard *Lacerta agilis*: about 8in=200mm., tail forming more than half. Sandy heaths and dunes with heather.

{ 9 Slowworm *Anguis fragilis*: male.
{10 Female slowworm.
{11 Young slowworm.
Well grown adult measures 15in=380mm., tail forming more than one half. Dry country of heaths and hedgerows, etc.

The small dust-coloured beetle climbs with pain
Over the small plantain leaf, a specious plain! . . .
The tender speckled moth her dancing seen,
The vaulting grasshopper of glossy green,
And all prolithic Summer's sporting train.

"Speckled moths" there are in plenty for both the bird's-foot trefoil *Lotus corniculatus* and the restharrow *Ononis repens* that grow here abundantly on the older dunes are important food plants for several moths and butterflies. The larvae of the common heath moth *Ematurga atomaria* feeds on bird's-foot trefoil as well as on heather. In most summers, though not in 1976, one of the commonest butterflies of the Sandling heaths and benthills is the common blue *Polyommatus icarus*. The colour of this delicate little butterfly is well described by South as blue with a tinge of violet in its composition, wings narrowly edged with black and veins a shining blue. It lays its eggs on bird's-foot trefoil and on restharrow, the caterpillar green with brownish hairs feeding throughout its life on these plants. On a cool Autumn evening of 1979 when the turf still retained the heat of the day, the adults could be seen in large numbers fluttering around the harebells or resting with wings furled for the night on the flowering stems of grass in the shelter of the dunes.

The chocolate brown hairy caterpillars of the fox moth *Macrothylacia rubi* which if touched sometimes cause a rash, also feed on bird's-foot trefoil as well as on bramble and heather, and can often be found here in late summer and autumn.

WASTE OF THE MANOR

In White's *Directory* of 1874 Sizewell is described as a hamlet of a thousand acres. This no doubt included some 400 acres of heathy lands adjoining Sizewell Common that fringe the low sandy cliffs and stretch inland to join Aldringham Common. Once extensive sheep walks and still known as the Crown Walks, they form part of the waste of the manor of Sizewell Hall, the former home of the Ogilvie family to whom the soil belongs.

These commons are criss-crossed by innumerable grassy tracks and bridle ways kept open in the past by sheep and shepherds, farm carts and horse-riders, gamekeepers, cottagers in pursuance of their common rights, fishermen walking to their boats and men and women walking to daily work in the opposite direction to Richard

Garrett's Engineering Works or to domestic work in the big houses.

For the past hundred years this area has been a pheasant shoot and game reserve and for the last half of the century at least, keepering with the squire's approval has been of the kind that allows local people to co-exist amicably with the pheasants with, at the time of writing, no aggressive keep-out notices to mar the peace of the place. Mr Fred Staff whose grandfather and great uncle were keepers and warreners before him, was born here on the common in Shell Pit Cottages. He remembers sheep grazing all over the Walks but he thinks it possible that some of the land was arable before it reverted to heath. This belief is based mainly on the keeper's log book kept by his grandfather from 1884–1907, where most of the early entries are for partridges which feed chiefly on arable. Pheasants began to be significant after the conifer and birch woods were planted in the 1920s when the annual count rose to 1000–1500 birds a year. The entry for rabbits taken in 1884 was 688 rising to 2,047 by 1887 and the total of rabbits shot between 1884–1907 was around 31,023.

In the direction of Sizewell Hall, seedlings from a fine old hawthorn hedge have developed into some thirty spreading hawthorn trees that add the scent of may blossom to the more typical heathland scents. Elderly but infinitely graceful birch copses *Betula pendula* give way to gorse and bracken where the heath becomes more open. The taller gorse *Ulex europaeus* seems to like disturbed areas on the verge of commons where man has been active. Weiss (1908) found from experiments that ants eat the seeds of gorse for the sake of their orange coloured oil bodies which they bite as they push the seed along the ground. As the ants often make use of human pathways, they drop some of the seed which, falling on the churned up edges of the track, take root and grow. In spring and summer and indeed most of the year, these ant-sown gorse hedges are packed with shining golden flowers so that on hot days the air is full of their almond scent.

Further on the heath where the eye is gladdened by a fair stretch of mature *Calluna* heather, the taller gorse gives way to low growing tussocks of western gorse *Ulex gallii*. Despite its name, this western gorse flourishes on some of our East Anglian coastal heaths where there are acid soils. It is in flower in late August and September with the heather when most of the taller gorse is taking a brief pause in flowering. Seen together in the field, the two gorses show their different characteristics: the shorter and usually more compact growth of the western gorse, sometimes no higher than the heather clumps, with narrow flowers and smaller spines, curved and scarcely furrowed, compared with the thick set bushes of *Ulex europaeus* with larger, deeper furrowed spines and flowers a deep yellow.

Like other members of the pea family, gorse bears on its roots little bumps or nodules carrying their own supply of nitrogen in the form of nitrogen fixing bacteria. Acid soils are often deficient in nitrogen so this ability to create nitrogen enrichment in the soil around the roots makes gorse a useful heathland plant.

Country people have many associations particularly with the taller furze or whin as it is variously called. In times when farm wages have been low and coal dear, furze faggots have been an important source of fuel. In Suffolk the tall bake ovens were stuffed with furze and commercially gorse was used to fire brick and lime kilns. Wind proof and stock proof, it had innumerable uses on farming estates where it served as raw material to construct whin-breaks, whin-huts, game coverts and along with hawthorn it was used in bush draining. Crushed green whins were widely used as fodder for cattle. Thomas Bewick recalls:

> In the early spring it was a common job for me before setting off to school to rise betimes in the morning . . . and equipt with an apron, an old dyking mitten and a sharpened broken sickle to set off among the whin bushes, which were near to hand to cut off last years' sprouts. These were laid in a corner till the evening when I stript and fell to work to "cree" them with a "mell" in a stone trough till the tops of the whins were beaten to the constituency of soft, wet grass and with this mess I fed the horses before I went to bed . . . It agreed so well with them with a little oats that they soon became sleek.

In one area of the Walks where the gorse has been burnt, new young heather, mostly *Calluna vulgaris* is flourishing and here on the sandy, acid soil grow associated plants like the heath wood-rush *Luzula multiflora*, the tall, downy heath groundsel *Senecio sylvaticus* and heath bedstraw *Galium saxatile*.

In contrast to this typically acid heathland flora is the vegetation on

127

what was once a single line railway running over Crown Walks via Thorpeness Halt to Aldeburgh and back to Leiston and Saxmundham. Where the material came from to build the track is not known. Local pits of crag or marl may have been used or, less likely, material may have been transported by sea. But whatever was used, plants are found along this tract that are more often found on chalk or limestone than on acid heaths. These include the carline thistle *Carlina vulgaris*, particularly striking in winter with its stiff structure and attractive colouring of straw and yellow gold; and the greater knapweed *Centaurea scabiosa*, both plants of chalk and limestone.

The bare trodden sand and grit on the surface of the old line on parts is red with mossy stonecrop *Crassula tillaea*, a tiny annual with minute oval leaves and thread-like stems that turn red in the sun as described by McClintock and Fitter (1967). Two rare little clovers grow on the Walks. One of these is suffocated clover *Trifolium suffocatum* which carries its flowers almost on the ground so that they appear "suffocated" by the taller leaves. The other is clustered clover *Trifolium glomeratum*, a prostrate, hairless little annual with small pinkish-purple round unstalked heads clustering along the stem. Both are described as very local in dry, bare and grassy places.

Another interesting small annual of dry barren places found along the track is fern grass *Catapodium rigidum*, little more than two inches high with narrow purplish leaves and one-sided rigid flowering heads with stiff fern-like ascending branches. Other plants of the barer patches include little mouse-ear *Cerastium semidecandrum*, small

cudweed *Filago minima*, and small cat's tail grass *Phleum bertolonii*. In early March sunshine on the bare sand the short-lived common whitlow grass *Erophila verna* is already opening deeply cleft white petals raised only an inch or two above the ground on single leafless stems, later bearing the distinctive flat, roundish seed-pods on slender stalks. So called ephemerals like the whitlow grass manage to dodge most of the summer drought by completing their life cycles early in the year.

COMMON WHITLOW GRASS
EROPHILA VERNA

Plants of dry heaths and windswept places are equipped with a variety of devices to prevent excessive loss of water

128

vapour. The leaves of the heather family for instance are small with a thick cuticle permanently rolled so that the underside where the breathing pores are placed is almost entirely enclosed. In this enclosed space, moist air is trapped and the leaf surface is shaded from sun and wind. Some plants like gorse and broom reduce the actual surface of their leaves into needles so that there is less area for evaporation. Other plants, particularly on windswept coastal heaths and dunes, protect themselves against the drying out action of sun and wind by growing in rosette form, their leaves pressed close against the soil so that only the upper surface is exposed. The rosette keeps the soil immediately round the stem cool while the leaves themselves are in contact with any moisture held in the soil.

A RETROSPECTIVE LOOK AT SOME BIRDS OF THE SANDLING HEATHS

Times change and today naturalists shoot with the camera rather than with the gun but Fergus Menteith Ogilvie (1861–1918) youngest son of Alexander and Margaret Ogilvie of Sizewell Hall, Leiston, Suffolk, was described by his publisher Henry Balfour as a "sportsman naturalist of the best type". Claude B. Ticehurst (1932) in *History of the Birds of Suffolk* selects him as one of nine former distinguished Suffolk ornithologists and refers to his posthumus publication *Field observations of British Birds* as full of original observations. Bird studies on the heaths and commons that lie along the coast between Aldeburgh and Dunwich in the neighbourhood of Sizewell common where Ogilvie lived as a boy form the basis of the book. Historically it is of considerable interest in giving a clear picture of these Sandling heaths at the beginning of this century and the end of the last, showing how vegetation changes since that time are reflected in changes in bird population. Height of heather, availability of gorse, extent of grazed open heath, presence of rabbit holes—all these are factors that along with other considerations of climate and human pressure can affect the numbers of certain typical heathland birds.

At a time when sheep were still grazing the Walks of all these heaths and commercial forestry had hardly begun, F. M. Ogilvie was writing:

. . . Much of the land in the vicinity of the coast, consists of rough, heathy moorlands, intermixed with arable and marsh lands.

On some of these commons, gorse and bracken predominate; others are almost entirely covered with heather and little else, and are more suggestive of a Yorkshire grouse-moor than a Suffolk *Partridge-ground*. In size they vary from patches of a few acres to large areas some miles in width or length. The surface is flat or gently undulating, but in some cases, e.g. the Dunwich or Westleton "Walks", the ground becomes much more irregular and hilly.

Of trees there are none, and with the exception of a few stray thorn-bushes, there is nothing to be seen over the wide expanse of greater elevation except a bramble or a furze, which may reach a height of six or eight feet.

Much of the land fringing the moor was at one time broken up by the plough and brought under cultivation. The soil however was so poor consisting only of light sand and innumerable stones, that the experiment, in many cases proved far from successful, and much of the reclaimed land was allowed to fall back into its primitive condition.

In some cases this became quickly reclothed with furze and heather, and hardly recognisable from the surrounding common, except, perhaps for the remains of old banks which marked the boundaries of former fields; in others again, the reclaimed patches have remained naked and stony wastes, bare of any vegetation, except coarse grass or an occasional bracken. This condition is accounted for by the continual scouring of the light sandy soils by winds, and consequent removal of any seeds lying on the surface.

Here and there a patch of this kind 40 to 50 acres in extent may be found in the centre of the big commons, remaining but little altered since the plough was last over it.

In other parts of Norfolk and Suffolk, these lands are called brecks . . . the moorlands I am dealing with here possess an avifauna very similar to [this] Breck region. The birds of a region such as this are characteristic and well defined and in some cases restricted to the common-lands. For since there are no woods or trees or hedgerows, a vast number of the passerine birds, warblers and so forth are absent. As there is no water, wading and aquatic birds are missing and as there is no grain, the game birds are not found here in any quantity, except on the edge of the moor, where the arable land meets the heathery waste.

Ogilvie then divides the characteristic breeding birds of these heaths and commons into those that breed on the ground, under the ground in burrows (wheatear and stock dove), or content themselves with low bushes: Of the passerine birds

> Larks, buntings and pipits breed in abundance. The moorlands are the stronghold of our three British Chats—the Wheatear, Stonechat and Whin-chat . . . It is here that the rare and generally retiring Dartford Warbler can be observed. Nightjars are numerous and form a very characteristic feature of the bird life.

Of the more unusual raptors attracted by rabbits, he reports that in winter peregrine falcons are sometimes plentiful, while on rare occasions a white-tailed eagle puts in an appearance.* Rough-legged buzzards and hen-harriers, as they do today, visited in winter and the kestrel then as now could be seen all the year round hovering over the heaths, head to wind, scanning the ground below for its varied prey— small birds, short-tailed voles, long-tailed field mice, beetles, earthworms, lizards.

Ogilvie writes with affection of another resident of the furze commons and heaths which he terms "that charming little bird the Stonechat *Saxicola torquata*". Though fewer in number, the stonechat is still a familiar companion on the coastal commons between Thorpeness and Minsmere and in the cold winter weather of 1978 on the benthills above the shore, a pair was behaving exactly as Ogilvie describes them in the equally severe winter of 1894/5 when apparently the ground was frozen hard for eight weeks. He describes how

> . . . a walk along the Bentlings fringing the coast would reveal five or six pairs, serenely happy in their wintry surroundings; full of life and movement, boldly following the intruder from bush to bush flirting their short tails over their backs, and scolding incessantly, until they have driven him out of their compound.

He had a theory that stonechats (which nest under gorse bushes or in heather)

> . . . loved the poor commons best i.e. those on which there are poor rights and which are in consequence, rather closely cropped by the common holders for litter and kindling and so forth.

(Sizewell and Aldringham are both quoted as "poor commons".) This observation seems not unconnected with David Lack's findings (1933) when examining changes in bird distribution during

---

*It did several times in the severe weather of January 1982.

WHINCHAT

the early stages of forestry plantation on Breckland heaths. He found that birds seemed to be very much affected by the height of vegetation, more so than by its nature. Stonechats and whinchats, absent in very bare heaths, were equally common in vegetation of a similar height whether bracken or young pines, using the top as perches to sing and to make aerial flights after food, but when that vegetation grew beyond a certain height the chats decreased and disappeared.

The changes since the beginning of this century in the Sandling heaths are perhaps most strongly underlined by the status of one of the most specialised heathland birds—the stone curlew. F. M. Ogilvie writes of the stone curlew or Norfolk plover as he often calls it:

> They are essentially birds of the Steppe—like the Bustard frequenting open treeless wastes. To my mind one of the most interesting British Birds and unhappily one of those species that is in danger of extinction at no uncertain date. On my own part of Suffolk between Aldeburgh and Dunwich in 1908 they were increasing but only because they had been strictly protected, birds and eggs alike.

In his diary for 1898 he was recording Norfolk plover as unusually numerous at Sizewell with flocks of 15–18 seen several nights on the "Black Heth" and in 1908, he was still recording "Norfolk plover unusually numerous with flocks of 15–20 at Scotts Hall and several sited at Sizewell on the Broom Cover and Crown Farm Common."

Just after the last war, autumn assemblies of sixty to seventy stone curlews in a gravel pit at Minsmere were noted by H. E Axell but with the encroachment of scrub, agriculture and forestry onto the "open treeless wastes" the stone curlew has all but disappeared from the Sandling heaths. However, in 1978 the Rare Birds Panel estimated four pairs breeding in the coastal belt of Suffolk.

The type of low scrub heath favoured by the Dartford warbler is very different from the natural habitat of stone curlews but it also has disappeared entirely from these heaths for at least a half century. In his collection of local heathland birds, once housed in his museum still standing at Sizewell but now in Ipswich Museum, Ogilvie includes three male Dartford warblers, two taken on Westleton Walks and one on Sizewell Common June 13 1890. (The last record for Dartford warblers breeding in the area was 1927.)

The news is perhaps a little better concerning two other interesting birds in the Ogilvie collection—the red-backed shrike and the great grey shrike. The coastal belt between Aldeburgh and Dunwich was until recently one of the best localities for the red-backed shrike with twenty-seven pairs recorded in 1977–78 and fourteen pairs estimated for the following year. (*British Birds*: Vol 73 No. 1) but even here the birds have been doing badly of late. Recent cold wet springs may be the cause of the reduced breeding success of this summer visitor from dry continental climates which in Britain is on the north west fringe of its range. This may have been the reason for the disappearance of a pair that had for several years been our very close neighbours on Leiston Common, rearing young in a bramble tangle on the edge of the close-cropped heath of the study area, secluded from view by the garden wall. Dung beetles are the commonest item in this butcher-bird's larder, together with bumblebees, wasps, cinnabar moths' larvae, shrews, field mice and occasional fledglings of linnet, yellowhammers and partridges—all items that in a good season this dry sunny heath supplies in abundance.

The great grey shrike is a regular winter visitor to these coastal heaths and on December 15th 1979, a male landed on a dead balsam poplar a few feet from the house. Here it remained a full five minutes turning its head this way and that giving a fine display of its contrasting black, grey and white plumage, with long black and white tail fanned in classical shrike manner.

While linnets, yellowhammers and meadow pipits continue as in F. M. Ogilvie's time to be the most frequent residents of the commons, the pine plantations have introduced the green wood-pecker as a constant visitor to the heaths for food, its yaffling call a sharp reminder of its presence. With crimson head and yellow-green mantle gleaming in the sun, it comes in search of ants in the short grass or scales the telegraph pole to probe the cracks for insects with its long tongue.

IMMATURE CUCKOO

In the immediate neighbourhood of the study area of Leiston Common, increase in garden bushes has attracted other birds, some less typical of heathland such as the willow warbler and the chaffinch: blue, coal, long-tailed and great tit; common and lesser whitethroat; hedge sparrow, redpoll and spotted fly-catcher. As in Ogilvie's time, the wet common and marshes, now with a small reservoir and drainage stream, attract various ducks, water-rail, moorhen and coot as well as reed and sedge warblers. Either the nests of the sedge warblers or those of the meadow pipits on the dry heath act as host for the egg of the cuckoo that visits us each spring, waking the common early with its persistent call.

A young cuckoo in its rufous stage imprisoned itself in the fruit cage on the edge of the heathland study area. While disentangling it from the net, close examination was possible of the red-brown barred mantle and striped whitish fawn underparts. Ogilvie found these immature cuckoos fed mainly on hairy caterpillars with some small beetles. The stomachs of two he caught in Suffolk in September and October were crammed full of the heads and empty skins of the yellow and black downy larvae of the buff-tip moth *Phalera bucephala*. This is a moth with silvery grey and violet colouring and rounded buff tip that frequently comes to light on Leiston Common, causing astonishment by its similarity in repose to the broken twig of silver birch.

NIGHTJARS OF THE SANDLING HEATHS

One of the most positive developments in the last half of this century has been the protection afforded to the Sandling heathlands by the nature conservation movement, with designation of these important coastal heaths as Sites of Special Scientific Interest and part of the Suffolk Heritage Coast.

One of the breeding birds to benefit from this development has been the European nightjar or fern owl *Caprimulgus europaeus* that Ogilvie referred to as being a characteristic feature of the bird-life of the heaths at the end of the last century when one of their favourite haunts were the commons around his childhood home at Sizewell. Nightjars, though in decline, still breed on most of the Sandling heaths but it is on two nature reserves that their requirements have been specially studied.

When nearly ten years ago an agreement was arrived at with the Home Office for some fifty acres of Hollesley Heath to be managed as a local nature reserve by the Suffolk Trust for Nature Conservation, the land was described as an area of mature Sandlings pinewood bordered on the south and east by formerly open heathland showing various stages in succession to birch and pine scrub. The intention was to create zones of all representative types of vegetation from bare ground to climax woodland, with the proportion of two-third open

NIGHTJAR

135

heath to one-third closed woodland. Careful management since has succeeded in maintaining a habitat that suits the breeding require-ments of nightjars.

It is a magical experience around 9 p.m. on a still June evening, in the company of the warden, Michael Cavanagh who knows these birds intimately, to witness the twilight activity of the nightjars. The glade where we wait is typical nightjar country with some young heather interspersed with patches of bare ground for suitable nesting sites where trees have been felled on the fringe of birch and pine copse. At first nothing breaks the hushed silence of the place. Then, suddenly, like a great moth, a male flies quite close as though to investigate us—so close that one gets a good view of the white tipped outer tail-feathers and the three spots on the wing tips, the wide beak and perfect "dead leaf" camouflage of the speckled buff and brown plumage.

Now the evening social life of the nightjars is about to start. First come the soft nasal flight calls followed by the loud snapping sound of "wing clapping" as first one and then another bird flies in and out of vision against the shadowy stage set of the trees. Exactly how the sound is made by the wings no one knows for sure. Only considerably later the characteristic churring sound is heard as the males return temporarily to their roost. In the deepening dusk others answer and the strange vibrating chorus comes now from all around.

It seems that this churring is a way of claiming territory. Nightjars are very sociable birds. Though it is said they rarely churr in each other's territory, the males go visiting each other in their respective territories early in the evening. Wing clapping and the flight call are part of this social exchange. The birds feed on many kinds of moths associated with various trees but not necessarily with open heathland, so they travel far into fields and oakwoods and are not reliant on food within their own territories. At the end of the evening, as we found, the birds go quiet. This is the time when communal feeding takes place in neutral zones outside the territories. Feeding often continues until just before dawn when the birds return to the regular roosts at the end of the night's activity.

Some forty years and more after David Lack's work (1932) on their breeding habits, Rob Berry when assistant warden of Minsmere Bird Reserve made a special study of nightjars from 1975–78 as part of a longer project. I am indebted to him for sharing his experience in a conversation that began with the subject of adders as two nightjar eggs had just been found with a double puncture—a sign that an adder had sucked them. The cold summer of 1978 had meant that the adders were very numerous on the heaths early in the year where they had stayed up longer than usual in the dry areas. As he explained the eggs are laid in a very shallow scrape usually to one side of a clearing. Nightjars are good parents, remaining on the nest until the last moment when approached. Their chief defence is feigning injury but this unfortunately makes no impression on adders! Not that they alone are to blame for losses. Jays, foxes, weasels, stoats all take the eggs which are laid on bare ground and so are very vulnerable.

Past and present studies show that the nightjar is a highly specialised bird with a self-destructive habitat. It nests on the fringe of silver birch trees *Betula pendula* that are encroaching on open heather-covered heathland. It rarely breeds in dense woodland so as the thicket advances, it advances with it. A typical nesting site is a small clearing with some bare ground among heather *Calluna*, near a sheltering tree averaging some three metres in height.

Britain is the only place where nightjars are associated specially with heathland and here their decline is closely linked with the general encroachment of birch since myxomatosis, though human interference and cold springs resulting in single brood production may also be factors.

At Minsmere, the average territory per pair is about 5.5 hectares (13½ acres) and in 1981 eight pairs bred at three different sites.

A more detailed report of Rob Berry's research is available in his article "Nightjar habitat and breeding in East Anglia" which was published in *British Birds* for May 1979.

OTHER HEATHLAND NATURE RESERVES OF THE SANDLINGS

Over two hundred hectares of acid heathland form part of Minsmere Bird Reserve (R. S. P. B.) Here the glacial sands and gravels overlying the Westleton Beds (see plate) give rise to these entrancing heaths, with silver birch as the dominant tree species. It was the plentiful supply of gravel that made possible the construction in the wetlands of what has become known as "the Scrape" which helped to encourage the nesting there of the avocets and terns for which the reserve is famous.

In contrast to the important gains in new wetland breeding species, Herbert Axell (1977) states:

Most of the losses have been from the heathland despite the work we have done there, due to changes since myxomatosis, with tall, rank growths of ling, heather, birch and bracken . . . In twenty years, eight pairs of stonechats reduced to 4, whinchat reduced from 15 pair to one, yellow wagtails and woodlarks ceased to nest in the area, red-backed shrikes reduced from 14 pairs to two and most regrettable of all—stone curlews ceased entirely to breed.

(Stone curlews used to breed among the sand and glinting flint of a forty-acre field, long ago colonised by seedlings of a sixty-year-old Scots pine plantation on the outer perimeter).

Attempts were made to recreate the nesting requirements of wheatears by sinking wood and metal boxes to simulate rabbit holes after myxomatosis reduced the rabbits but this had only temporary effect. The result was the same when stone curlew territory was recreated by rotovation of quiet parts of the heath. Clearly, as Herbert Axell suggests, some other essential factors were missing that the birds required.

Adjoining Minsmere are about 100 hectares (250 acres) of the heathland of Dunwich Common, previously owned by Dunwich Town Trust but since 1968 the property of the National Trust. Former rights by certain freemen of Dunwich to graze sheep and cut bracken on Dunwich Common seem to have lapsed entirely but public footpaths at the time of writing are well maintained and very fully used.

Long years of devoted land management by the warden, J. C. Docwra, including the shifting of hundreds of tons of pebbly soil, has achieved some interesting ling-covered heathland. Ninety per cent of the ground cover is *Calluna vulgaris* but at the cliff's edge cross-leaved heath *Erica tetralix*, more often a wet heath plant, grows on apparently dry soil among the dominant heather. Here, too, in August it is possible to find the small pale pink flowers of dodder *Cuscuta epithymum*, a semi-parasitic plant that trails its stems like a fine red network around the stems of heather cross-leaved heath and gorse, its host plants. Western gorse *Ulex gallii* is on the extreme east of its range on this East Suffolk coastal heath and is increasing rapidly. It grows little higher than the heather and on one of those golden September days that this coast is heir to, when both the low western gorse and the heather is in flower, alternating mauve and yellow hummocks spread wave upon wave to the cliff's edge in a brilliant setting of sea and sky.

Yellowhammers, linnets and meadow pipits, as one would expect, abound on these coastal heaths. Because of the birds and other wild life including the glow-worms *Lampyris noctiluca*, of which sixty-three were recorded during one count in 1979, no spray is used. Bracken and birch saplings are coped with by hand with the help of local conservationists and volunteers from schools, or by the swipe.

Beyond the heathlands cared for by the Royal Society for the Protection of Birds and the National Trust are 47 hectares (117 acres) of Westleton National Nature Reserve, one of the few extensive open tracts of heather *Calluna vulgaris* left, some of which has reverted to birch woodland. Adjoining this land are 445 hectares (1100 acres) of the Walberswick National Nature Reserve of which about half is heathland.

Sheep have walked these lands since mediaeval times leaving their traces in such names as Westleton Walks, East Walk, West Walk and Sheep-wash Lane. To provide more of the open grazed ground of former times a small section of land is fenced off with a view to introducing some experimental sheep-grazing.

Bracken today is rarely harvested so as the fronds die, the rachides decay at ground level and the shading effect finally kills the heather. While the Nature Conservancy Council Reserve Warden, C. S. Waller would prefer biological control, it looks as though here as elsewhere the world-wide problem of bracken spread must be treated in a variety of ways. Asulum, a spray thought to be harmful only to bracken and perhaps sorrel, has so far proved somewhat disappointing but trials continue, together with ploughing and rotovating as well as cutting with a forage harvester which is proving useful as it removes a percentage of the thick litter and protective covering, leaving the bracken more vulnerable.

Fierce accidental heath fires can increase the growth of bracken with its protected underground food storage equipment at the expense of heather, but planned controlled burning of heather has historically been used to produce vigorous young growth and keep down scrub. Complex questions are involved, however, and since the decline of rough grazing on southern lowland heaths, the controlled burning at special times of the year that used to accompany it has also declined as a management tool.

Many people now believe that heather is self-regenerating and does not begin to die back until thirty years or more. Mature heather i.e. *Calluna vulgaris* that is twenty to thirty years old, is a valuable and very rare habitat, so that for several reasons the value of burning is under constant review. Thirty-three species of insects, for instance, are known to feed on heather while in the thin litter of fallen leaves under short heather on some open ground heaths in the south west, Delany (1956) found fifty species of animals of which thirty-eight were mites, mostly orabatid mites that are associated with decomposing plant material and themselves play a role in the decomposition process as part of the economy of the heath. Under heather clumps it is much cooler than on bare ground in hot sun and here such creatures as the woodlouse *Porcellio scaber* like to live during the day. Although these creatures living in the soil and litter may not be killed by fire, Chapman (1978) found that they decline rapidly after a burn when there is no plant cover to stop the litter drying out or being scattered by the wind.

Where controlled burning is used on nature reserves its aim is to provide a regular supply of heather of different heights and ages since many birds, reptiles and other creatures dependent on heather have special preferences in this matter. With this in view an experiment in the controlled burning of a mosaic of small plots at graded intervals has been carried out on Westleton Nature Reserve over some fifteen years so as to ascertain the best intervals at which to burn in order to achieve an age succession. The conclusion will of course depend on the end in view, but a twenty-year interval seems likely to be most satisfactory for heather production and least damaging to wild life. Partly as a result of these experiments, Westleton is considered to have some of the best managed heather stretches of southern heaths.

The Walberswick National Nature Reserve is nationally important for its fresh-water reed beds, home of the bearded tit and probably one of the largest uninterrupted areas of its kind in Britain. Where temporary pools occurred on the sandy coastal heaths towards Southwold there were records in the 1920s of natterjack toads *Bufo calamita*, still found further north at Winterton but no longer found in east Suffolk.

D. J. Pearson (1973) describes the changes in the bird population on the Walberswick heaths from 1953–1972 with the advent of myxomatosis in 1954. He mentions the spread of silver birch with over a hundred acres of bracken giving way to birch woodland in fifteen years. He records the general decline of specialised heathland birds, with wryneck, wheatear, and woodlark having disappeared completely; stonechat and redpoll increased but otherwise only the tree pipit maintaining its numbers. He suggests the decline is probably connected with long term climate changes and widespread habitat destruction in south-east England. The overriding importance of the habitat, however, is perhaps proved by the glad news from the warden that in the last three years, management of the heaths so as to provide more short heath on open ground has succeeded in enticing back in small numbers wheatear, woodlark and whinchat—all short turf feeders.

It is clearly scientifically important that heathland habitats should continue to exist in order to ensure the survival of plants and creatures that depend on them. It is also clearly important for our own quality of life that at the same time people should be able fully to appreciate and enjoy these green open spaces without in any way feeling excluded from a heritage our ancestors fought so hard to keep.

This is the importance of a little reserve near Aldeburgh set aside for educational purposes where environmental advisors, Nature Reserve and Heritage Coast wardens and teachers combine to give schools a close acquaintance with nature in a heathland area. An R.S.P.B. reserve, North Warren as its name suggests is a rabbit grazed area of grass heath with open ground of the type referred to by F. M. Ogilvie as infertile sandy ground, once ploughed and then, because of its poverty, allowed to revert to heath. If the open breck-like ground, attractive to stone curlew and wheatear in the past, can be maintained (and rabbits are certainly once more doing their best), with the River Hundred, a reed bed, alder scrub and old railway track to bring variety, it should prove a rewarding field of study.

When large numbers of young people become personally involved in this way, there is real hope for the future of Nature Conservation and the heathland habitat.

Concern for East Suffolk's remaining heathlands outside the reserves has recently brought together local government and conservation groups and the public with a view to preventing the deterioration of habitat by the invasion of bracken, birch and pine. A vegetation survey particularly of heathland which is also common land is proposed in order to determine the ecological quality of the heaths, with the Tunstall/Blaxhall area being used as a pilot scheme for practical work.

Measurements quoted indicate the *average* length from bill-tip to tail-tip.

1 Kestrel *Falco tinnunculus* (13½in=343mm): Resident, passage migrant and winter visitor to Suffolk heaths esp. near coasts, farm-lands, open woodlands.

2 Dartford Warbler *Sylvia undata* (5in=127mm): A former resident no longer found in Suffolk.

3 Linnet *Carduelis cannabina* (5¼in=133mm): Resident, winter visitor, passage migrant. Breeds on most Suffolk heaths.

4 Red Grouse *Lagopus lagopus* (15–16in=380–405mm. Female smaller than male): Ticehurst (1932) describes how a few pairs turned out on Breckland early in the century and given an artificial water supply, increased to 150 pairs by 1908 but did not survive the war. Essentially a bird of upland moors and peat-bogs where it breeds near heather or cranberry in Scotland, N. England, Wales and the border countries of Shropshire, Hereford and Monmouth, and in Ireland. In south-west England, grouse were introduced in the 19th century where they are still found on Exmoor and Dartmoor. Female smaller than male.

5 Golden Plover *Pluvialis apricaria* (11in=280mm): Winter visitor and passage migrant in Suffolk. At Sizewell, Suffolk in Dec. 1906 before heavy snow Ticehurst recorded an unusually large flock of 3–4000 birds. Breeds from southern Pennines in N. E. Yorkshire northwards, in Ireland and more locally in Wales and on Dartmoor, Devon.

6 & 7 Hen Harrier *Circus cyaneus* (17–20in=432–508mm. Female larger than male): Now breeds chiefly on northern moors among thick heather or rushes but in the past also bred elsewhere i.e. on Suffolk heaths at the beginning of the century (W. H. Payn 1978). Today annual winter visitor particularly to east Suffolk coastal heaths and marshes. Female larger than male.

8 Stone Curlew *Burhinus oedicnemus* (16in=405mm): Scarce local summer visitor, occasional in winter. Breeds very locally.

9 Lapwing *Vanellus vanellus* (12in=305mm): Breeds decreasingly both on cultivated land and on coastal commons and the Breck. Resident, winter visitor and passage migrant in Suffolk.

10 Ring Ouzel *Turdus torquatus* (9½in=240mm): Spring and autumn passage migrant to Suffolk in small numbers. Breeds usually above 1000ft among heather/rocks in most counties in northern and western England west of a line from Yorkshire to Devon. Also in Wales.

11 Sky Lark *Alauda arvensis* (7in=178mm): Widespread resident, winter visitor and passage migrant in Suffolk. Breeds in open country.

12 Wheatear *Oenanthe oenanthe* (5¾in=146mm): Local summer visitor and passage migrant. Breeds locally.

13 Yellowhammer *Emberiza citrinella* (6½in=165mm): Common resident, winter visitor and passage migrant. Breeds widely on Suffolk heaths.

14 Stonechat *Saxicola torquata* (5in=127mm): A local breeding bird on Suffolk heaths, winter visitor and passage migrant.

See W. H. Payn (1978) *The Birds of Suffolk*; BOU (1971) *Status of Birds in Britain and Ireland*; C. B. Ticehurst (1932) *A History of the Birds of Suffolk*. Measurements: R. Peterson et al. (1974) *Birds of Britain and Europe*.

Stretching some thousand square kilometres across the borders of Norfolk and Suffolk in the heart of East Anglia is the Breckland—a strangely haunting land with a quality all its own, whose essence W. G. Clarke in the first quarter of this century captured incomparably in his book *In Breckland Wilds* (1925). In his time it was one of the largest areas of open heathland in lowland Britain, boasting at least seventy-four named heaths and warrens. Behind the figures given in White's *Directory of Suffolk* for 1874, lies a graphic picture of how the western part must have looked in his day:

> *Lakenheath* comprising 10,550 acres of land, including a large portion of fen on the west and an extensive tract of light sandy land on the east and south, including a rabbit warren of 24,000 acres . . . *Brandon* well built market town noted for gun flints, whiting, rabbit skins and furs pleasantly situated on the south bank of Little Ouse river, which is navigable for barges . . . 6760 acres including 4500 acres of light sandy soil which was enclosed under an act passed in 1807 previous to which it was open sheep walk and large rabbit warren. Though enclosed still many rabbits. It borders the extensive rabbit warren of Lakenheath, Santon Downham and Elvedon which supply Brandon furriers with immense quantities of skins, dressing of which gives employment to 200 females . . . *Santon Downham* sandy parish . . . on the south side of the Little Ouse. Remarkable for the invasion of sand which in 1668 threatened to overwhelm the whole parish burying several houses and cottages and choking the navigation of the river . . .

Thetford, Breckland's main town, was capital of East Anglia in Saxon times. Long before that, as we have seen, neolithic farming people were attracted to Breckland because unlike the heavy clay areas it was probably not covered with dense deciduous forest, nor was it swampy like the Fens. The presence of local flint meant there was plenty of raw material from which to make tools so that, by evolving a type of shifting agriculture that abandoned plots of land as they became exhausted, it was possible to support quite a large population. The Domesday returns seem to suggest that there were only one or two hundred acres of woodland left in Breckland at the time of the Norman conquest. It was at this time that the grazing effects of sheep, goats and cattle were intensified by the introduction of rabbits which probably increased erosion and the creation of heathland on the sandy soil by preventing the regeneration of woodland. During mediaeval times, Breckland became the most famous area in England for rabbit warrening.

Before the seventeenth century, unlike the early enclosed central Suffolk, only limited enclosures had taken place in the Breck, mostly because the lords of the manor needed a clear run for their very large flocks of demesne sheep. They had no wish to upset the traditional foldcourse system which before the introduction of turnips as a fodder crop, they shared with other sheep-corn producing areas of Norfolk. In this system heathland played a vital part in the economy.

A typical manorial estate on the Breckland sandy soils of north-west Suffolk and south-east Norfolk at this time consisted of a large permanently cultivated open-field area with great outlying stretches of heaths and commons, patches of which were broken up by the plough, cultivated for a short while then allowed to revert to heathland. Such areas became known as brecks and it was this traditional practice and the special quality it gave to the land that led Clarke to give it its now generally accepted name of Breckland. In most Breckland villages manorial influence was strong. The village lands including the heathlands, were usually divided between two or three large manors, the land allotted to each being known as a foldcourse in which every cultivator, landlord and tenant, farmed a varying number of strips scattered over compact sections of land in the open fields set aside as *shifts* for the planned cultivation of winter and spring cereals. Sheep were important and if these were to have pasture for the whole year, it was essential for each foldcourse to include both open arable field and sheepwalk since the latter was important for summer feeding when the arable was under crops. After the fields were harvested, then sheep and cattle could be put on the harvested fields for *shackage* as feeding off the stubble was called.

Both landlords and tenants enjoyed important common rights over the open fields and heaths. Generally speaking the lord's sheep fed all over the unsown arable lands, the demesne heaths, the harvest shack fields and also, though this was not part of the foldcourse, the commons. The tenants used the commons and fed their great cattle on the harvest field after harvest, and also put a limited number of sheep in the lord's flock all the year round.

During the seventeenth century land patterns began to change in ways welcomed by the young progressive farmers but lamented by others. Nathaniel Bloomfield in his poem concerning *The Enclosure*

146

of Honnington Green reflects the different viewpoints in the countryside around 1800:

> And tho' Age may still talk of the Green
> Of the Heath and free Commons of Yore
> Youth shall joy in the new fangled scene
> And boast of the change we deplore.

Some heathland brecks, particularly those near the arable lands
which had first been manured by sheep and then temporarily
ploughed up, began to be enclosed and cultivated on a more
permanent basis—a process that increased with the Parliamentary
Enclosure Acts of the eighteenth and nineteenth centuries. Nevertheless, M. S. Postgate (1962) to whom I am indebted for this
information, considers that as late as 1843 at least 8000 acres of
commonland still remained in thirteen parishes of Central Breckland.

Today, the landscape has greatly changed since Francois de la
Rochefoucauld wrote after journeying from Bury St Edmunds to
Thetford: "The whole country through which the road runs for a
distance of 8 miles is covered with heather—not a shrub, not a plant
except in the little valleys one can see in the distance . . ." This picture
of great tracts of open heathland, chalk grassland and inland sand-
dunes has been greatly modified by the planting on a large scale of
soft wood conifers. The Forestry Commission began planting in the
early 1920s and they have now over 50,000 acres (about 20,250
hectares) getting on for quarter of the total landmass of Breckland
(*Forestry Commission Guide* 1972) Agriculture, too, by heavy liming
and the use of lucerne in a rotation of crops, has reclaimed a fair
proportion of the original heathlands. However, if one searches,
enough survives to keep alive the spirit of the place thanks to the
development in this century of Nature Conservation and the skilled
and devoted team work of wardens and their voluntary helpers.

ITS UNIQUE QUALITY

What is it about Breckland that leaves one with that longing to
return—"that hunger for the heathland that must be satisfied," as
Clarke put it? I remember the unique quality of the Breck being
discussed during a day tour to the area as part of the British
Ecological Society's Symposium of 1970 in which heathland ecology
featured largely. Expert opinion seems to be agreed that the primary
cause of the strikingly individual character of the area is the

combination of light porous soil with a semi-continental climate. Breckland has, at an annual average of about 600mm, one of the lowest rainfalls of any inland area in Britain. Summers are normally hot and dry and winters cold with pockets of frost occurring in every month of the year.

The whole region rests on a foundation of chalk. Some patches of this chalk lie on the surface but practically all of Breckland is overlaid with drift deposits of various depths. These are a mixture of chalk pebbles, crushed flints and sandy gravels of various origins with hardly any heavy clay. Over all is a blanket of wind blown sand, lying deepest on low lying sites and plateaux. Chemically these soils range from highly calcareous shallow young soils with chalk just below the surface to older soils where the sand is deep and very acid, at times of the humus/podsol type.

As one would expect, the natural communities of plants and animals vary according to the soil sometimes within quite short distances, but three of the dominant plants seem able to grow both on the thin chalky soils and the deep acid sands. This is true of the grasses *Agrostis/Festuca* ssp., of the sand sedge *Carex arenaria* and of bracken *Pteridium aquilinum*, but these only flourish and become dominant on the soils that really suit them. For instance, areas of loose sand are thickly colonised by the sand sedge that is only sparse in other areas.

Heather *Calluna vulgaris* that covers acres of Breckland heaths where the soil is sandy and acid, hardly grows at all on the highly calcareous areas. This is shown in the photograph (see plate) of a heath near Weeting. Here the dark stripes in the vegetation are caused by the growth of heather along troughs of acidic sand in otherwise chalky soil which *Calluna*, a calcifuge plant, is unable to tolerate. This visible pattern of alternative ling and grass corresponds to a similar underlying pattern in the soil, which is composed of "soil stripes", a phenomenon of the area whose causes go back ultimately to frost effects in the very cold conditions of the last Ice Age.

All the plants mentioned above are often described as highly social and aggressive in the sense of growing in close knit communities and spreading at each other's expense, and many studies have been made by E. P. Farrow (1925), A. S. Watt (1955) and others to try to find under what circumstances, for instance, bracken will invade and finally shade out heather. Some factors involved are the age of the communities, the character of the soil and, very important, resistance to various degrees of grazing.

148

Although there are basic differences, Breckland and the Sandlings of coastal Suffolk have some things in common. In the blown top-soil of the Breck grow some plants that are usually only seen near the coast. Sand sedge *Carex arenaria* flourishes in recently drifted sand as it does on our coastal heaths and the dune grey hair-grass *Corynephorus canescens* that is found on the coast at Minsmere, in Breckland occurs miles from the sea on the inland mobile dune system of Wangford Warren. Sand too, attracts the ringed plover that nests here in a scrape in the sand as it does on the coast. Both regions share the dryness of soil, the high number of hours of sunshine, the quality of light and some of the lowest rainfall in Britain so that in open plant communities or where overgrazing keeps vegetation low and soil nearly bare, the same little winter annuals have a chance to ripen their seed and survive. One recognises in the short turf and rabbit scratched sand the tiny spring-flowering plants like the early forget-me-nots *Myosotis ramosissima* that for a brief spell in their thousands carpet the ground and then are gone.

Later flowering are the rosy patches of mossy stonecrop *Crassula tillaea*, so familiar on the barer ground of certain parts of the Sandling heaths, which in Breckland also grows on the well trodden paths and tracks of Cavenham Heath and elsewhere. Walking over the heather-covered dry heaths of the Breck one notices, however, the almost total absence of bell heather *Erica cinerea* which occurs frequently on the Sandling heaths with their less extreme climate and nearness to the sea.

AN ANCIENT SHEEP WALK

I first heard the call of the stone curlew on Weeting Heath—a wild and eerie sound in a wild and eerie place. This Breckland heath is now a nationally important nature reserve and one of the main tasks of the warden is keeping an eye on this rare heathland bird whose breeding haunts in Britain are largely concentrated in the few remaining steppe-like regions of the Breck.

According to sixteenth-century records Weeting dry heath was historically used as a sheep walk. It was open commonland until the enclosures of 1775 and much of it has probably never been cultivated. In the early nineteenth century, it became part of a large shooting estate and it was then that groups of Scots pine *Pinus sylvestris* began to be planted as shelter for game. In the first half of the present century it was stocked with rabbits and some 2000 acres (810 hectares) were used for rabbit farming. During the last war it was used by the military until 1942 when it was given by J. C. Cadbury Esq. to the Norfolk Naturalist Trust, which now has a Reserve agreement with the Nature Conservancy Council.

Ten years ago in September of 1969, Weeting was a starting point for a personal visit to Breckland planned with a view to seeing certain areas in detail, by kind permission of the Nature Conservancy Council and the Norfolk Naturalist Trust.* On a day of brilliant sunshine in a week of continuous downpour we approached the land adjoining the Weeting Rabbit Enclosure on the south of the reserve. Bounded on two sides by shelter belts of Scots pine *Pinus sylvestris*, the soil showed little sign of the gallons of rain it must have received. Around the enclosure (erected in 1959), small mesh wire netting encircled forty acres of ground in an experiment that sought to reconstruct conditions that were typical of the area before myxomatosis—calcareous stony grass heath, very open, sparsely covered with close-cropped vegetation and riddled with rabbit holes: the type of land that is favoured by the wheatear and by the stone curlew, as well as some special Breckland plants for whose survival the scratched soil surfaces as well as the short turf seems important.

The first impression was of a wide treeless stretch of land covered with flints and large round stones. Suddenly, however, some of the rounded "stones" bounced away and one could see that the warren was alive with rabbits. According to the warden's information at the time there were as many as thirty-five rabbits to the acre, a total for the enclosure of 1400, about half of which could be above ground at any one time. Outside the enclosure rabbit density was of course very much reduced. The effect on the vegetation was obvious. The soil consists of sand over chalk, thin in some places but several feet deep in others. Large areas of the enclosure were covered with mosses and lichen interspersed very sparsely with mainly sheep's fescue *Festuca ovina* grass and in places sand sedge thickly colonised loose sand.

Where rabbits disturbed the soil, above all it was ragwort *Senecio jacobaea* that flourished, its yellow flowers here and there providing colour among the black and white of flints and grey-green of moss and lichen. So thin was the leached soil that moles *Talpa europaea* tunnelled almost on the surface in search of the chrysalids of the vivid

---

* Because of their sensitive character, permits are required for all Breckland sites apart from the open access area on Cavenham Heath.

Cinnabar Moth *Tyria jacobaeae*. The larvae of this moth, with its vermilion and charcoal colouring, are the yellow and black striped caterpillars that feed on ragwort, in some years stripping their host plant of every particle of green. Interaction of cinnabar moths and their food plant was studied on Weeting Heath from 1966–77 by research scientists J. P. Dempster and K. H. Lakhani from the Monks Wood Experimental Station of the Natural Environment Research Council. The abundance of the moth fluctuates violently and it periodically defoliates the ragwort over large areas. The plants react to defoliation by increased vegetative reproduction from root buds resulting in undersized rosette flowers that do not flower. While the study showed that it was rainfall in the preceding summer that determined ragwort abundance in terms of plant numbers, it also showed that the caterpillars determined what proportion of plants flowered and produced seed. In years such as 1976 when defoliation was severe, no seed at all was produced. The resulting reduction in the bulk of food available can lead to high mortality in the larval stage of the moth from starvation, and those caterpillars that do survive produce small adults that lay fewer eggs the following year.

LARVAE OF CINNABAR MOTH
TYRIA JACOBAEAE
ON RAGWORT

A RETURN VISIT

Returning to Weeting Heath ten years later in July of 1979, after a late spring and prolonged rainfall, the general appearance of the heath was more green and fertile, due partly to an unusually wet spring, partly to a decrease in rabbits.

It was a delight on this visit to see several special Breckland plants in flower. On the dry grassy verges of the brecks were the small yellow flowers of sickle medick *Medicago falcata* and the later flowering field wormwood *Artemisia campestris*. Further on the heath, among the turf where the sand is thin over the chalk the dainty maiden pink *Dianthus deltoides* was flourishing and spreading, with here and there the small white flowers of squinancywort *Asperula cyanchica*

SPANISH CATCHFLY *SILENE OTITES* AND VIPER'S BUGLOSS MOTH *HADENA IRREGULARIS* WS 32 mm

nestling in the grass. Elsewhere, on the sloping sides of an old marl pit of the kind found all over Breckland, three species of thyme grow including the special Breckland thyme *Thymus serpyllum*.

Two plants are very local indeed and rarely found outside Breckland: the Spanish catchfly and the spiked speedwell. Of these the Breckland or Spanish catchfly *Silene otites* is one of the so-called "continental" species and is not found elsewhere in Britain. Although it is stickily hairy and catches insects in the same way as the rest of the catchfly family, it looks very different with its rosette of dark green leaves and delicate spikes of pale green flowers. This is a plant that depends on open and disturbed ground in which seeds can germinate and it prefers shallow chalky soil with low vegetation. The Spanish catchfly is the food plant of the viper's bugloss moth *Hadena irregularis* so it follows that this moth is found only in districts where its food plant grows. South (1977) says that it was first found in the Breck and district in July 1868 on viper's bugloss *Echium vulgare*—hence its misleading common name. It was soon found, however that its caterpillar fed on *Silene otites*.

The second very local plant that was in flower on that memorable July day was the beautiful spiked speedwell *Veronica spicata*. This rare perennial once flowered freely on open grass heath in Breckland when there was heavy rabbit grazing, but many of the heaths where it grew have been ploughed up and since myxomatosis in many remaining sites it has been unable to grow through the litter of tall grass and other plants now growing unchecked. It grows in sandy soils and spreads by stolons. In Breckland, small colonies of spiked speedwell still survive and are shielded from too much grazing during the flowering and growing season though at other times light grazing is considered essential. In late July of 1979, the long spikes of flower heads were packed with purple flowerets.

Other plants almost confined to Breckland are spring speedwell *Veronica verna*, Breckland speedwell *Veronica praecox*, drooping brome grass *Bromus tectorum* and perennial knawel *Scleranthus perennis*. (Trist 1979)

WHEATEARS AND STONE CURLEWS

The bare rabbit-cropped heaths and sheepwalks that suit these Breckland plants also suit the wheatears and the stone curlews. Wheatears were once one of the commonest summer visitors to Breckland but with the spread of fir woods and growth of bracken since myxomatosis their numbers have decreased dramatically though Weeting Warren is still a favourite haunt where they can be seen feeding on the stony ground, white rumps showing as they flit from place to place.

Even more affected by change on land use has been the stone curlew, a bird whose appeal is as compelling and elusive as the Breck itself. A warden who lives in the field, as it were, with the stone curlews gains a rare insight into the habits of these birds of deserts and wide open spaces. I am told that at times there is an extraordinary stillness about the birds. Male and female will gaze at each other for long stretches of time as though nothing were taking place behind those large yellow vacuous eyes. Then suddenly, simultaneously and with great intent they set about some joint activity. The Rev. W. Gilbert White of Selborne wrote:

> Oedicnemus (Thick knees) is a most expressive name for them since their legs seem swelled like those of a Gouty Man, yet they run with the swiftness of a Greyhound, and sometimes stop suddenly, holding the head and body motionless . . . The young run immediately from the Egg like Partridges, and are withdrawn to some flinty field by the Dam, where they skulk among our grey spotted Flints, which are so exactly of their colour, as to be such a security that unless he catches the Eye of the young bird, the most accurate Observer may be deceived.

I am indebted to the present N. C. C. Senior Breckland Reserves Warden Martin Musgrave for the following interesting comments on the relevance today of Claude Ticehurst's well-known account of the birds (1932). Ticehurst says that the stone curlews return to the same spot every year, nesting in a scrape on open, barer parts of the heaths, on cultivated land that has gone back to waste, on the stony part of Breckland's poor grassland and sometimes on fallow, often depositing their two eggs in a lining of rabbit's droppings. The sites chosen today are still often associated with rabbits i.e. short cropped vegetation with flint strewn bare ground. Egg markings are very variable but often individual and characteristic of each female. Both male and female share in incubation and the egg shells are normally eaten by the parent birds after hatching to avoid discovery.

Stone curlews are most active at dusk and dawn when they can make a considerable clamour. Ticehurst says that their note changes after the eggs are laid but Martin Musgrave says that he has heard their call, best described as a sort of "musical scream" throughout the breeding season. Some feeding takes place during the day, but it is generally thought that feeding activity increases at night.

Stone curlews are particularly prone to disturbance and this of course limits the number of sites where they nest. According to a recent survey there were twenty-eight nests in Norfolk and thirty-five in Suffolk in 1974.

SOME EXPERIMENTS IN MANAGEMENT

Because of the national need to increase the type of vanishing habitat that is required for the continued existence of Breckland's special birds, insects and plants, several experiments are at present in progress. One of these has the stone curlews in mind. Recently new wide firebreaks and rotovated plots were made to include outer swathes of freshly turned earth with flints and these are already showing signs of attracting the birds. Another experiment attempts to reconstruct a traditional breck on what is known as the Breckland Weed Reserve. It consists of a field that the farmer has agreed with the Nature Conservancy Council to cultivate in the traditional manner using kidney vetch *Anthyllis vulneraria* as a nitrogenous green manure crop with less insecticides and fertilisers. A third rather different trial is being carried out on newly-acquired arable land on Roper's Heath where an attempt is being made to speed up reversion to heathland by cropping land without feeding it thus exhausting the nutrients and producing the impoverished condition of the soil that seems to give the heather communities a start over other plants. After harvesting in the normal way, the plot will be rotovated then allowed to do its own thing. Careful records will be kept as to which wild plants are first to arrive and which finally take over.

As the relict heathland areas become smaller, more fragmented and disturbed, the plants and animals are increasingly endangered, so this gives added importance to Lakenheath's 510 hectares (1260 acres) of the largest remaining area of unspoilt heathland in the Breck. This land, adjoining the American air base, is now a Site of Special Scientific Interest Grade 1 as described in the *Nature Conservation Review*, but it remains part of the Elvedon Estate owned by Lord Iveagh. Lakenheath is strictly private as are all Suffolk Breck sites owned by the Elvedon Estate.

Besides possessing large areas of *Calluna*, this is the only site in Breckland showing the complete sequence of grassland soils referred to in a famous series of studies by A. S. Watt who from 1931–36 recorded how the vegetation varied over quite short distances according to soil types. All soils were then sheep and rabbit grazed and varied from highly calcareous shallow soils, to neutral loams, to deep acid podsols. With the gradual change from base rich to more acid soil, plants ranged accordingly from such species as Moonwort *Botrychium lunaria* and the carline thistle *Carlina vulgaris* that thrive on chalky soil to plants of acid podsols such as heath bedstraw *Galium saxatile*, heath woodrush *Luzula campestris*, shepherd's cress *Teesdalia nudicaulis* and various *Cladonia* species of lichen. (More tolerant plants of course overlapped a range of soil types.)

On Cavenham Heath in April of 1979 important experiments were in progress in the management of bracken and heather. Here the serried ranks of the advancing front of bracken are moving more rapidly than they did ten years ago and a five-year programme is underway for bracken control. Various methods are being tested including rotovation, mowing and, where suitable, spraying with the chemical, Asulam. Answers are being sought as to the most effective seasons and number of times to cut bracken and whether it is better to leave the fronds or remove them.

Regarding the heather *Calluna vulgaris*, as on all other reserves, the aim is to produce varying stands to suit the varying preferences of

155

creatures that make their home on heathland. From the point of view of the renewal of the heather plant itself, it is of course essential to encourage plenty of young growth. Here again tests are in progress to compare the results of mowing versus rotovating with either no grazing, partial or permanent grazing by rabbits and roe deer and perhaps the re-introduction of sheep.

HEATHER BEETLES AND TIGER BEETLES

Some fine areas of young heather are being produced to take the place of the many stands of old heather dying of old age. Unfortunately however, in the spring of 1979 the heather beetle *Lochmaea suturalis* had been at work in exceptional numbers. After a winter of high rainfall these little brown heathland leaf-beetles, which on grouse moors are devoured by grouse, in the south and east have proved exceptionally prolific and in April were crawling all over the heather including the young heather in the experimental areas, leaving the tips with a singed brown appearance. It is the young tips that the larva feed on and they can spoil, if not kill, large areas. In its turn this leaf-beetle is fed upon by the nocturnal violet ground beetle *Carabus violaceus* and also *Carabus problematicus*.

A more welcome inhabitant of hot, sandy heaths was also around in fair numbers in April of 1979. These were the tiger-beetles *Cicindela campestris* that could be seen flying around in short rapid flights over the young heather and the short turf of heathland paths, their colours of varying brilliance in the noon sunlight. One of these, the elegant jade creature in the coloured illustration, survived captivity well while under observation.

The tiger-beetles are fierce hunters, beautifully designed for the part: long slender copper-coloured legs for fast running, very prominent eyes, and formidable mandibles that when closed overlap in two great curved blades. Both adults and larvae are meat-eaters. A vertical burrow about a foot deep is excavated in the sandy soil where the larva lies in wait for its prey, hooked in position by two hooks in its back. Its curiously bent body is described by A. D. Imms (1947) as being a perfect fit to enable it to hold onto the walls with its head blocking the entrance, jaws nicely placed to snap up passers-by.

CONTRASTING HABITAT

Breckland has been described as East Anglia's desert and its desert quality is certainly part of its fascination and its unique quality. I was, however, struck on my last visit by its variety of habitat. Where heathland merges with the alluvial flood plains of the River Lark, sluices and newly excavated dykes are gradually returning the water-meadows to the condition they were in before water levels some years ago were drastically lowered by the Water Authority. Best foil of all to the hot dry heaths are the fens of Chippenham National Nature Reserve where excellent care has produced some of the best fen meadows in the country with a succession of orchids throughout the summer, unforgettable in mid-July when the fragrant orchids *Gymnadenia conopsea* in abundant flower are filling the air with their scent.

Part of the heaths yet with their quite separate identity are the heathland meres with names that go back to Saxon times—Ringmere, Langmere, Fowlmere and the Devil's Punch Bowl. No stream flows in or out of these ancient meres for they are fed directly out of the chalk beneath, so that they are deepest in summer after winter rains have seeped far into the soil, but occasionally liable to dry out altogether later in the year when a summer drought has lowered the ground table. Beyond Langmere there is a fine shimmering vista of uninterrupted stretches of wavy hair-grass *Deschampsia flexuosa*, one of the heathland grasses that have benefited from the decrease in grazing. Near here the Drove Road crosses the heath. Once the meres were important watering places for the cattle and their drivers who travelled hundreds of miles along this historic track. Today it is the roe deer that come down to the meres to drink along with weasels, stoats, grass-snakes, adders, toads and innumerable other creatures of the heathland community.

THE SURREY HEATHS REVISITED

"Here I am at Thursley after as interesting a day as ever spent in my life. They say that *variety* is charming and this day I have had of scenes and soils variety indeed."

Cobbett's words, written in his journal of Nov 24th 1821 after one of his Rural Rides in the county of his birth, recalls just such a splendid day spent in the same area with the late F. L. Reynolds, Conservation Officer for the Surrey Naturalists' Trust, some ten years ago. The extent of heathland has shrunk since Cobbett's day but it is still a wonder to find some 14,000 acres (5666 hectares) of heath and common so close and accessible to the metropolis with some of the most exhilarating views anywhere in the country.

Variety of soil there is in plenty for the commons are of three principal types: the Greensand heaths especially around Hindhead where there are still probably three or four thousand acres of open or scrub heath; the heaths of the Tertiary Sands including Chobham, Esher and Wisley commons; and the less extensive commons on the plateau gravels above the chalk.

The sunshine of May 26, 1970 was at midsummer heat but the greenery of the Surrey countryside had still the freshness of spring. It was a good time to see Chobham Common, a Site of Special Scientific Interest of national importance with its outstanding insect and spider fauna. It is owned by the Surrey County Council and spreads over some 1700 acres (690 hectares) of the Tertiary Sands of the Thames. Some seventy million years ago, after the chalk had been laid down, massive earth movements caused upheavals in south-east England creating the great dome of the Weald and the complementary hollows of the Hampshire and London Basins. Later a shallow sea (the Tertiary Sea) filled the basins, depositing in the London Basin layers of clay and sand over the chalk. One of the uppermost layers consisted of extremely coarse infertile Bagshot Sand. This covers part of the Thames Basin and gives rise to some light oakwood but mostly heathland with birch and pine, of which Chobham Common is typical.

Like most of these Surrey heaths, the common until the last war had been grazed from time immemorial by sheep, cattle or ponies, but of recent years, through lack of grazing, many areas showed signs of fast regenerating to scrub and finally woodland with self-seeded birch, pine and some oak saplings. There were, however, at the time of my visit some ten years ago, considerable stands of mature heather (*Calluna vulgaris*), with its varied associations of plant, animal and insect life a valuable and increasingly rare habitat.

Crossing all the Surrey commons are innumerable ancient rights of way trodden in the past by cottagers cutting turf and peat and kept open by the horse-carts of the "broom-squires"—families who until the last war lived on the common and made their living by selling brooms fashioned from heather and birch. These ancient rights of public access were safeguarded for posterity by the application of the 1925 Fresh Air and Exercise Act, and exist even on the extensive lands used by the Ministry of Defence except at Westend Common and

158

Colony Bog, a site of Special Scientific Interest where because of danger from unexploded ammunition the site had been entirely closed.

It was near Aldershot on one of the heaths of the Lower Greensand that I received a memorable insight into the complexities of managing an area where so many interests and ownerships exist side by side, from the ever-present army to the horse riders, naturalists, golfers and the rambling public. We lunched on a slope among the roots of old pines, above us the typical birch, heather and gorse combination of dry heath. Below in the hollow, chance damming of a little stream had created a boggy area around a stretch of open water and here was the magical sight of several square feet covered with flowering shoots of bogbean *Menyanthes trifoliata*, the manifold mauve blooms rising from a lush carpet of the plant's leaves. Near it, too, was growing the rare slender cotton grass *Eriophorum gracile* only found in a few very wet bogs on southern heaths.

Maintaining the delicate balance of this small area required considerable skill based on the painstaking understanding of the subtle requirements of all its forms of life. There were, for instance, the dragonflies in the preservation of which the Surrey heaths have played a significant part. Yet the whole habitat owed its existence in the first place to the chance breaking of the ford which had somehow raised the comparative level of the paths and so prevented water flowing away at speed through the gulley on the other side of the road. I remember F. L. Reynolds saying that he was keeping his fingers crossed that no tidy minded officer would come along and order its repair, totally oblivious that he was doing anything but good. In some areas of the reserves Ministry of Defence boundaries cut right across habitats where soil water levels are crucial and either draining or flooding could cause catastropic changes in vegetation.

Sunlit parachutes gliding over the heights of Hankley provided the first pleasant impression of this common with its dry open heathland. Here we were on the Greensand with superb views stretching to the Hindhead Ridge on the far horizon, and the route taken by William Cobbett when following the Wey valley from Aldershot at the end of the last century. The land was then described by him as "*bare*, sandy heath". Certainly the term no longer generally applies. On the May morning in question, the light green of young birch and the blue green of the pines mingled with the darker stretches of heather.

Bare loose sand in which to lay eggs and some mature heather is the combination needed by sand lizards which are now only found in one or two isolated groups on the West Surrey heaths. The Greensand heaths were an important site for another scarce and disappearing reptile, the smooth snake (*Coronella austriaca*). These snakes may still be around but their wide ranging habits make them difficult to trace and the Herpetological Society has found no reliable proof of their presence on Surrey heaths in recent years.

A COUNTRY PARK

One place where sand lizards continue to thrive is Frensham Heights. Today a National Trust Property administered by the District Council, Frensham Common is a very ancient common where human presence stretches back to mesolithic times, and the recorded history of Great and Little Pond to at least the thirteenth century. For many city dwellers it has long provided a lifeline to fresh air and exercise and ever-increasing pressures from human recreational needs make it an outstanding illustration of the contradiction that lies in the heart of conservation. The conservation movement in the long run can only succeed if the public is behind it. Yet those very people who come in vast numbers to an area like this because they love and need it, can unwittingly destroy the very thing they want most to survive. Ten years ago, the erosion round Frensham Great Pond bore eloquent witness to this fact.

Revisiting Frensham in 1979, there was immense improvement. The common had been designated a Country Park, thus making grants available for the provision of two wardens. Much of the bare sand in the eroded areas had been reclaimed both by seeding and the scattering of heather cuttings. Where uncontrolled horse-riding was becoming a menace to people and environment, signed horse-rides have now helped to solve the problem. Above all there are plans to totally reshape the car park, scrapping the present one and site it in woodland well clear of the pond where those who wish to sit in their cars and gaze out at the water can do so in pleasant circumstances with no threat to the environment.

In the midsummer of 1979, the focal point of a return visit to the Surrey heathland reserves was Thursley where William Cobbett ended his ride. The Thursley/Elstead/Ockley group of Royal Commons cover some 13,000 acres (5260 hectares) with a range of habitat from dry heath to bog, supporting a varied range of plants and animals. It was a pleasure therefore to find that this important Site of Special Scientific Interest had been bought from its private owners by the Nature Conservancy Council and given the status of National Nature Reserve.

It is above all for their dragonflies that the Thursley complex of commons is famous. Twenty-six of the forty British species occur in the dozen or so heathland bogs and pools at Thursley making it the most important site for these insects in the country.

Among those breeding in the pools, some of the most successful species during the four years covered by a survey made by F. Reynolds (1969–73) were dragonflies that are particularly dependent on the type of habitat the Surrey heaths have to offer—marshy land and peaty bogs with heather. Among these is the slender small red

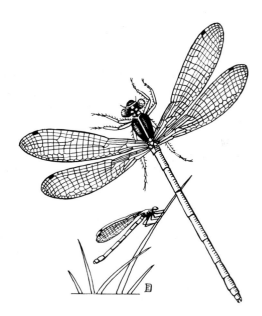

SMALL RED DAMSELFLY
CERIAGRION TENELLUM
MALE WINGSPAN 36 mm

161

damselfly *Ceriagrion tenellum* confined to southern counties and very local even there because it breeds only in boggy ground where there is heather over which it flies in feeble flight. Also flourishing here as breeding species were three darter dragonflies of the family Libellulidae: the four-spotted chaser *Libellula quadrimaculata*, the black darter *Sympetrum scoticum* and the keeled skimmer *Orthetrum coerulescens* with its distinctive powder blue colouring, all dragonflies which favour marshy ground, peat bogs and peaty pools to breed in, with heather as a resting place.

(Successful breeding species of course also include those of wider habitat preferences like the large red damselfly *Pyrrhosoma nymphula* and the common blue damselfly *Enallagma cyathigerum* of the family Coenagridae, as well as the larger dragonflies such as the brown hawker *Aeshna grandis*, a dragonfly of open ground of the family Aeshnidae.)

The rarest and nationally most important of the Surrey dragonflies is the white-faced dragonfly *Leucorrhinia dubia*, Surrey being the only place south of Chester where it breeds. This scarce little darter dragonfly, with its shining white face, black body and red or yellow spots, breeds in peaty bogs especially where there is heather nearby. According to Cynthia Longfield (1949), it is an alpine species of the Pyrenees and the Alps extending north east into Siberia. The nymph lives in peaty pools and sphagnum swamps on heather moors, where "the adult flies low over the heather with a rapid dodging sparrow-like flight" frequently returning to a chosen perch low down in the heather.

Wet heath and sphagnum-covered valley bog cover a considerable area of the Thursley reserve. The bogs are generated by water sinking through sand from higher ground and hitting a flat impervious layer of Greensand with some clay, causing a horizontal spread. Wooden planks transverse an area of quaking bog where the bog has developed by accumulated growth over and into open water, the peat and the moss having so little solidity that here unwary horse and rider have become engulfed. From this spot in mid-July one looks across a golden sheet of flowering bog asphodel *Narthecium ossifragum* splashed with the white fluffy fruit of cottongrass *Eriophorum angustifolium*. Sankey's Pool near at hand is fast filling up with vegetation but other pools being steep-sided do not clog in the same way. Dragonfly larvae feed on the maggots of small flies and beetles living in the pools but they also feed on each other. Surveys show that the water contains few beetle larvae but is crammed with dragonfly eggs and larvae so it seems likely that the young of the dragonfly feed mostly on each other.

On the fringe of the bog, on what may be termed damp heath, all three common species of heather grow, *Calluna vulgaris, Erica cinerea and Erica tetralix* as well as gorse *Ulex europaeus* and birch saplings *Betula pendula*. Among the gorse young stonechats in the sun were noisily making their presence known, and on the bark of a silver birch, a newly-emerged grayling *Hipparchia semele* resting with folded wings, allowed us to approach within a foot with only a fluttering of its eye spot as a warning to come no further. Nightjars breed on these heaths on the fringe of birch copses where there is medium height heather and some bare ground. The hobby on the other hand, now a regular breeding bird on Surrey heaths, chooses the edge of evergreen or mixed woods to nest in, building the nest high up in dark pines overlooking the open commons. Here the female can sit on the nest watching the male hunting over the ground, chasing insects and quartering the ponds for dragonflies. It seizes the emergent dragonflies with its claws and then transfers them to its beak, letting the discarded wings flutter down to the water below.

Hopes are being fulfilled that the new type fire-breaks of the current fire fighting improvements will favour the silver-studded blue butterflies. These breaks are so managed that there is always a central stretch of bare sand from six to seven feet wide surrounded on each side by young heather making a total width of thirty feet. It was a delight to see large numbers of silver-studded blues *Plebejus argus* loitering over the young heather either side of the paths as we walked, both the elegant blue males and the sooty brown females. The caterpillar feeds mainly on areas of short calluna heath where the eggs, usually a pale purple shade, are laid.

HEATH FIRES IN SURREY 1976

It was while visiting the Surrey heaths for the second time with the present Conservation Officer Arthur Lindley that I gained a salutary insight into the devastating effects of accidental fires which in the year of drought 1976 assumed horrendous proportions. Their threat to the heathland ecology is graphically conveyed in an important report drawn up jointly by the Surrey Trust for Nature Conservation and the Royal Society for the Protection of Birds. Its contents cannot be too widely known.

163

The report stresses Britain's international responsibility for protecting the vanishing heathland habitat because, while our southern heaths have been shrinking at a considerable rate from the beginning of last century, the loss elsewhere in north-west Europe has been even more severe and this country now holds some of the best heaths of this type in the world. As the report points out fire has always been a factor in the history of heathland and it remains a convenient management tool by which old vegetation can be cleared and new growth encouraged. However, very hot deep burns or too frequent burning will completely destroy the characteristic vegetation cover so that it is replaced by grasses which support a different type of wild life. Moreover, many species of less mobile creatures—such as reptiles and many insects are killed by fires as they cannot escape them. Where the fire covers only a proportion of large areas, recolonisation from outside the burnt zone is fairly straightforward and rapid but with the modern isolation and fragmentation of heaths, the matter is more complicated. Many small heaths exist for their inhabitants as fortresses surrounded by hostile territory, and recolonisation after a widespread fire has proved to be very slow or impossible.

By mid-August 1976 a minimum of 4085 acres (1653 hectares) of heathland had been burned in Surrey out of an over all total of 14,000 acres (5666 hectares), with the four worst fires occurring in the four most important complexes of commons. Very many important species of plants were destroyed and prevented from flowering and seeding. Only one or two extensive stands of mature heather, the most important element of acid heathland, now remain in Surrey. Many species of birds were nesting at the time of the fire and thus their eggs or young were destroyed. Though it is difficult to quantify the effects of the fires, it is estimated that the habitat destroyed supported the following rarer bird populations: hobby 2 pairs, nightjars 45 pairs, Dartford warbler 3 pairs, woodlark 25 pairs, stonechat 40 pairs.

The full effect of the fires may be judged from details given in the report for the Thursley/Elstead commons. Here there was the added hazard of burning peat smouldering underground and bursting later, when unattended, into flame. Two fires occurred affecting three quarters of the heath, with extensive areas of gorse and heather destroyed. (Houses in the village were only saved by a firebreak cut with army bulldozers at the last minute.)

At the time many birds, including stonechat, nightjar and

woodlark as well as commoner species were incubating or feeding second broods; these were all killed. Two pairs of curlews (half the Surrey breeding population) lost their nests in the fire. An area of stunted pine usually favoured by breeding whinchat was destroyed. Silver-studded blue and grayling butterflies suffered particularly badly. Shortly before the fire there had been a big emergence of blues—more than had been seen for many years—and although some of these were able to get away from the fire thousands were killed. The graylings were still in the chrysalis stage, so every one in the area was burnt. One of the two established sites for sand lizards on the common was burnt, and although it was hoped that they might have survived in their burrows, none of these lizards have been seen on the site since the fire so it would seem that all were killed . . .

It is the loss of habitat that is the most serious consequence of fires. Because of the drought, much of the normally water-logged peat in the area was dry and therefore burnt away. This may affect the water holding capacity of the bog and mean that the specialised plants, insects and other creatures that depend on this type of habitat will no longer be able to live there. The most valuable heathland habitat of all is mature heather and all but one of several large stands of this were lost. It will take up to twenty-five years of new growth to replace it.

Sobering reading, carrying a responsibility that should weigh heavily on everyone who values our heathlands. There are, however one or two items of good news. Three years later there was the encouraging sight of a vigorous growth of young bell heather *Erica cinerea* and dwarf gorse *Ulex minor* in the dry heath area that had for two years after the 1976 fires remained nothing but bare charred earth. Luckily sufficient heather was saved near enough to allow of natural re-seeding. Both bell heather and dwarf gorse are the first to recover from fire but smaller seedlings of ling *Calluna vulgaris* were also growing well and in time would probably shade out and replace bell heather as the dominant plant.

The second bit of good news concerns woodlarks. These ground-nesting birds like low heathland vegetation and a certain amount of open space. They seem to have benefited on the whole from the clearance made by the fire, building up to nine pairs on Thursley by 1979.

A VISIT TO THE HEATHS OF DORSET AND THE NEW FOREST

Thomas Hardy visualised the successive stages in the seasonal cycle of change on his Egdon Heath as the green or fern period of the morning of the year, the flowering or noontide phase when the July sun fired the crimson heather to scarlet, and "the brown period when the heath bells and ferns would bear the russet of the evening; to be in turn displaced by the dark hues of the winter period representing night".

By this reckoning, my first visit to Furzebrook Research Station and the Dorset heathland Nature Reserves was at the beginning of the brown period in October 1969 followed some ten years later by a return visit in the green or young fern period of the heaths in May 1979.

Whatever the time of the year, the heathlands of the Isle of Purbeck and the Wareham district live up to their national reputation as some of the finest lowland heaths in Great Britain. The geographical limits of several heathland species pass through southern England. Here east and west meet so there is a wealth of interest to which one needs to return again and again, but between the two stays it was possible to see many of the plants and animals for which these heaths are important sites.

"VAST TRACT OF UNENCLOSED WILD"

Remembering how Hardy believed no one could understand the heaths who had not been there at sun-down, I was glad my first view of the fine stretch between Arne and Wareham was at that transitional point of time when the heaths begin their "nightly roll into darkness". This is the hour of greatest contrast between earth and sky when night settles down on the sombre heathland vegetation while daylight still lingers in the sky above, giving the scene that peculiar intensity at twilight described in *The Return of the Native* when "Looking upwards, a furze-cutter would have been inclined to finish his work; looking down he would have decided to finish his faggot and go home".

My home for the night was a cottage in Marnhull near to Tess of the D'Urberville's birthplace—an ancient village to the north-east of the beautiful Vale of Blackmoor, whose stone quarries have yielded proof of more or less continuous human occupation on the site from

600 B.C. In these quarries, along with neolithic arrow heads, the bones of small Celtic sheep have been found whose grazing thousands of years ago helped to preserve "the vast tract of unenclosed wild" referred to by Thomas Hardy as still in existence in 1895 when he was writing the introduction to the first edition of his novel. At that time, according to him, the original unity of the heaths was "somewhat disguised by intrusive strips and slices brought under the plough with varying degrees of success or planted to woodland". The fragmentation of the heathland had already begun.

The Dorset heaths overlay extremely infertile soils derived from deposits made up of white and yellow sands, local ironstones, gravels and some ball clay. These are the Bagshot Sands laid down in the Eocene period of geological time. The first Ordnance Survey Map of 1811 showed over 74,000 acres (about 30,000 hectares) of heathland on these poorer soils in east Dorset and in Hampshire west of the River Avon. During the next eighty-five years with the alternate boom and slump of agriculture, here as elsewhere heaths reclaimed by the plough reverted to heathland so that by the 1896 Ordnance Survey, at the time Hardy was writing, the general picture was not very different from the beginning of the century except for some urban development round Bournemouth. By 1960, however, according to Nature Conservancy figures, the heaths had declined to 10,000 hectares (24,710 acres) and in 1973 to about 6000 hectares (14,826 acres)—a fifth of the 1811 figure. As a result of these changes Hardy's vast tract of unenclosed wild has been fragmented into more than a hundred pieces.

Nevertheless, among the Dorset heaths are still to be found the greatest expanses of lowland heath remaining in these islands. Four of the scientifically most interesting are now heathland nature reserves and therefore safe from further encroachment. All are in the neighbourhood of Poole Harbour. Three of them—Morden Bog (149 hectares), Studland Heath (631 hectares) and Hartland Moor (258 hectares) are administered by the Nature Conservancy Council.* A fourth, Arne, is administered by the Royal Society for the Protection of Birds. Dorset has in all over 8000 acres (3240 hectares) of commonland, nearly all heathland according to the Royal Commission on Commonland.

A HEATHLAND RESEARCH STATION

On this warm coast, within easy reach of both the heaths of East

*In addition the NCC has just acquired Holton Heath (115 hectares) and Three Barrows (64 hectares).

Dorset and the New Forest is Furzebrook Research Station—a centre of ecological studies that can look back on over a quarter of a century of really exciting research into the heathland habitat.

Answers have been sought to a whole range of intriguing questions: Why does the beautiful Dorset heath *Erica ciliaris* which occurs only on a few sites outside the damp heath and peaty sites of Dorset occur abundantly within certain well-defined areas in the county but not at all on other sites that seem equally suitable? Why is it that the elimination of certain common species like heather by grazing or the rabbit by myxomatosis, has a radical effect on the whole heathland while the presence or absence of other species—all rare and many common ones—apparently has little comparable effect? After a heathland burn, what is the pattern of regrowth in heather *Calluna vulgaris* throughout its lifetime and what are the effects on its associated plant and animal life? Why is the rare large marsh grasshopper *Stethophyma grossum* less widespread than the bog bush cricket *Metrioptera brachyptera* when both like to live in much the same boggy areas? How much heather is eaten by the heather beetle *Lochmaea suturalis*, the emperor moth *Saturnia pavonia* and the fox moth *Macrothylacia rubi* during their life cycles? (In view of the damage done to heather in the wet summer of 1979, it is interesting to know that a population of heather beetles of 100 individuals per square metre would consume 250 grammes of heather per hectare during their life span, while each larva of the much less populous emperor moth would consume about 8 grammes dry weight.)

Furzebrook is now part of the Institute of Terrestial Ecology of the National Environmental Research Council. An excellent feature of this research establishment is the lively and accessible manner in which its scholarship is conveyed to the public. This has been the case since the early days of Furzebrook when it formed part of the Nature Conservancy whose first Director General, Cyril Diver, was its founder and, it is generally agreed, sowed the seeds of much of the research in progress today.

It was a south-west regional officer of the then Nature Conservancy, N. W. Moore, who in the early days of Furzebrook did pioneering work on the effects of changes in land use and consequent fragmentation of the heaths on heathland flora and fauna. He based his studies on ten species found on a series of Dorset and West Hampshire heaths of differing degrees of isolation. Five of these—the Dorset heath *Erica ciliaris*, the small red damselfly, *Ceriagrion tenellum*, the silver studded blue butterfly *Plebejus argus*, the

Dartford warbler *Sylvia undata* and the sand lizard *Lacerta agilis*—were specific to heathland while the remainder were also found on other habitats. Moore found that species more or less confined to heathland tended to die out altogether on the smaller, more isolated heaths. None of the ten species studied survived the plough up and fertilisation of heathland, though in the case of forestry some managed to adapt for a while in forest rides or early stages of plantation.

One of the two birds studied—the Dartford warbler—I saw for the first time here in Dorset, "popping up out of the gorse bush like a cork on the end of a long feather" as my guide M. V. Tuck, Warden of Hartland Moor and Morden Bog National Nature Reserve described it. It is now virtually restricted to the New Forest and Dorset heaths and the heaths of Surrey where Moore found its ideal habitat was high dense heather *Calluna* mixed with gorse *Ulex europaeus* on a south facing site. Gorse that has become open and leggy, however, ceases to be a suitable habitat. He argued that since furze is closely associated with human activity and moderate burning, cutting and grazing helped to produce the type of heath that the Dartford warbler enjoyed, it could be said that the presence of this warbler depended on the activity past and present of people. However, climate too is an important factor for the birds' numbers are greatly reduced in hard winters, and Dartford warblers are here on the northern fringe of their range.

More recently (1976) research biologist Colin Bibby by microscopic analysis of the birds' faeces has shown that the adult Dartford warblers live mainly on a diet of beetles, spiders, flies, bugs and caterpillars that feed on heather and gorse. Gorse is important because it flowers throughout most of the year so this means that stable food items like the gorse weevil *Apion ulicis* or the caterpillars of moths that lay their eggs in gorse flowers, are available all the year round when other food is in short supply. The young of the Dartford warbler, on the other hand, are apparently fed on bigger moths and caterpillars found on birch and other trees so in spring the parents must forage further away from the nest beyond the immediate scrub.

ANTS ON HEATHLAND

All this many angled research has confirmed the need of heathland management to aim at maintaining as wide a variety of heathland conditions as possible across a whole range of different aspects and moisture levels, since here lies the best chance of creating suitable living conditions for the widest range of plants and animals. This point has been particularly well illustrated by certain discoveries concerning ant distribution and behaviour—an aspect of Furze-brook's work that is among its most important contributions to heathland ecology.

Quite apart from their place in the food chain, ants have an effect on soil fertility and hold an important place among heathland invertebrates. They lick up juices that plants exude and, on heathland, collect nectar from heather flowers and from the nectaries of the young fronds of bracken.

On Hartland Moor, a knoll of dry heath rises out of wet heath and this for the past twenty-five years has been the scene of a series of experiments by M. V. Brian and his team to find out what ants live on the heath and whether different mini-habitats have a special appeal for different species, what were these ecological differences and how did they help to reduce mutual interference, enabling species to co-exist at close quarters.

Ants were attracted to sugar lump baits over a ten hectare area (slightly under 25 acres) and detailed records were kept of surrounding plant cover, moisture, organic matter content and temperature of soil etc. The greater portion of the site is what naturalists recognise as "warm, dry heath"—the sort of habitat that is known to attract ants in cool climates like our own. Strung out on a route starting at this warm dry end and stretching down towards the cold wetter heath near the bog, different ant species were found to have chosen to make their home in conditions most nearly approximating to those of their world geographical distribution.

On the dryest and least vegetated area was *Lasius alienus*, a Formicine ant of southern continental origin. In the dampest, coolest part of the heath was the black ant *Lasius niger*, a more northerly member of the same family better able to tolerate cold weather. In between the warmest zone sheltered from wind and not subject to wet was the territory chosen by the turf ant, *Tetramorium caespitum*, a Myrmicine ant of middle European origin well equipped to live on southern English heaths where it chooses the best parts of dry acid soil dominated by heather, dwarf gorse, bell heather and many associated plants.

It was found that soil moisture, vegetation height and underlying geology all affect ant distribution. At the same time, the differing abilities of each kind of ant enable them to cope with their particular

170

environment. *Lasius alienus* for instance, can cope with the extremes of climate in dry bare heathland because it cuts underground galleries in the bare soil so that it avoids strong winds and has the benefit of the direct warmth of the sun yet is shielded from too extreme heat. It thrives best in the early stages after the heath is burnt but suffers and tends to disappear when the heather grows and thickens. This is when neighbouring colonies of the turf ant expand their range for these ants can build mounds of soil in the vegetation which raise them up towards the sun's warmth as the heather grows. They also have the advantageous habit of collecting seeds, particularly of heather to store underground over winter for spring.

The ants' physical characteristics often fit them for their special niche in the environment. The small brown energetic workers of *Lasius alienus* that are designed to live underground in warm regions, have no hairs, whereas the black workers of *Lasius niger* are covered with dense hairs that help them to survive in cooler wetter areas where they live mainly above the surface in denser vegetation, making use of clumps of heath and grass tussocks to lift them above the water table. The turf ants *Tetramorium* tend to keep the two species of Lasius apart and their own adaptation to a different mini-habitat enables these two ants, with a similar diet of small insects and honey-dew, to co-exist. The ants mature their flying sexual stages at different times and this too aids their joint survival.

ANTS AND BUTTERFLIES

Recent ant research in Britain is helping to increase understanding of the life cycle and conservation needs of the large blue butterfly *Maculinea arion*, particularly through the studies of this butterfly by J. A. Thomas on its last British site.

Like many species of blue butterflies the larvae produce secretions that attract ants. The large blue chooses southern slopes where wild thyme *Thymus drucei* grows, usually with some gorse and in the vicinity of ants' nests. This is because the larvae require two food sources. At first they feed on the flowers of thyme then they leave these to wander about until they meet ants of the genus *Myrmica* that colonise dry shallow soils. M. V. Brian (1977) in his fascinating book on ants says that by the time the butterfly larvae are ready for the ants they are similar in size, texture and smell to the ant larvae. Whilst the ants are caressing it and licking its secretions, the caterpillar suddenly hunches itself and this appears to be a signal for the ants to take it

171

back to their nest where it spends the entire winter feeding on ant larvae until it finally pupates and emerges in its adult form in June.

When one lives on the bleaker coast of East Anglia, the desire to visit the heathlands of the south and west is naturally heightened by the chance of seeing plants and animals largely restricted to these warmer, more oceanic regions. In my case, for instance, there was the possibilty of seeing for the first time the marsh gentian *Gentiana pneumonanthe*, brown beak-sedge *Rhynchospora fusca*, Dorset heath *Erica ciliaris* and marsh clubmoss *Lycopodium inundatum* not to mention the sand lizard *Lacerta agilis*, smooth snake *Coronella austriaca*, the small red damselfly *Ceriagrion tenellum*, the large marsh grasshopper *Stethophyma grossum* and a pink bee-eating spider *Thomisus onustus*—all species for which Dorset is one of the main British sites. However, here as elsewhere it is not the rarities but the sum total of these marvellous heathlands that is the main attraction.

Studland Heath in a week of May sunshine after the drenching rains of the late spring of 1979 was a case in point. 8000 people use the dune route each year, many of them school parties and students, yet despite various signs of erosion here and there, the South Haven peninsula in May-time mood seemed to be taking it all in its stride. From the high barrow of Spur Heath, the smooth low contours of the land rolled away to the undulating sand dunes and the shore. The oldest, heather-covered ridge of dunes is around 350 years old. Since then the sea has piled up three more ridges of sand at intervals of roughly a hundred years. Cloud shadows today race across the peaty waters of Little Sea, once part of the open bay, now an inland freshwater lake entrapped in arms of sand as the peninsula grew seawards. The light and shade of the sun is echoed in the light and dark of the vegetation as colour changes mark the merging of duneland into heath, where the pale marram grass gives way to the dark bronzed green of the heather. Bell heather *Erica cinerea* establishes itself very early on the acidic soil of these dunes, decreasing inland as ling *Calluna vulgaris* becomes the dominant vegetation on dunes of about 150 years upward.

Meadow pipits are here in plenty and there are stonechats feeding their young but, perhaps because of the absence of grass dunelands and the dominance of heather, there are, strangely, no skylarks.

It was in the loose sand near the path over the heather-covered ridges of Studland Heath that we were lucky enough to get an excellent view of both male and female sand lizard, as well as a fleeting glance of a last year's youngster scampering into a hole in the bank. (For this sight as for so much else, we were indebted to Studland Reserve warden, Rees Cox.) The female was quite heavily built, her grey shape decorated its whole length with three lines of round, white-centred black patches. The green male, similarly patterned, was a splendid sight against the background of white sand and tall heather. Sand lizards *Lacerta agilis* like to lie in a sunny position like this, tilting their bodies to the sun and sunbasking. They live in colonies and burrow in the loose sand of open dry heaths, feeding on centipedes, woodlice, worms, spiders, moths and other small creatures. They are now only found on a few southern heaths.

All the British reptiles are found on Studland—even the grass snake *Natrix natrix* which is not really a heathland species but thrives in the wetlands near the woods that grow on the verge of the Little Sea—a place that also suits interesting plants like bog myrtle *Myrica gale*, bog-bean *Menyanthes trifoliata* and the royal fern *Osmunda regalis*. Here and around the crater pools, twenty species of dragonflies breed, among them the very local small red damselfly *Ceriagrion tenellum* can be seen from June onwards. The male's abdomen is entirely crimson red, the female's largely bronze, and both have red legs. This little damselfly is only found in the acid bogs and peaty runnels of Dorset, Surrey and a few other southern counties.

An afternoon spent lingering over alternating dry and damp areas of the dunes in full view of the sea ended at Spur Bog at the foot of Spur Heath where rivulets of peaty water seep under the road bringing with them a certain amount of calcium from road building operations so that the ground is enriched in places. Impeded drainage causes horizontal spread and waterlogged soil with here and there little runnels of open water. The hollows are filled with the yellowish green and umber of the bog mosses and growing among their lush hummocks are the rosy stickily hairy leaves of the round-leaved sundew *Drosera rotundifolia* and the less frequent oblong-leaved sundew *Drosera intermedia*.

Insectivorous plants like the sundews and butterworts solve the problem of poor nutrient levels in peat by supplementing what they obtain through their roots by trapping and digesting insects. When an insect lands on a plant, it is trapped by sticky glands that cover the leaf surface. The leaf then slowly wraps up the insect and takes from it the chemicals and minerals it needs.

Widespread on shallow peat as well as deeper peat is the downy grey-green foliage of the cross-leaved heath *Erica tetralix* with its whorls of four narrow evergreen leaves. Harvest mice surprisingly have nested in the untidy straw-coloured tussocks of purple moor-grass *Molinia caerulea* which later will send up the purple flowering spikes that give it its name. Here too in May the semi-parasitic lousewort *Pedicularis sylvatica* is bright with its pink two-lipped flowers and scattered everywhere in the wettest areas are the small sword-like leaves of the bog asphodel *Narthecium ossifragum*.

In a few places on the slightly raised peaty verges of the wet runnels one can find the flat basal rosette of the pale butterwort *Pinguicula lusitanica*, a variety of butterwort found only in the bogs and wet heaths of western Britain. The green, yellowish leaves roll inwards at the margins to trap insects on sticky hairs and later, in July, the plant will send up singly small lilac flowers with a pale yellow throat. These are smaller and paler than the deeper violet flowers of the common butterwort *Pinguicula vulgaris* that grows in similar habitats and is more widespread though scarce in the south.

Another neat little plant that grows here in short turf or on bare peat where there is little competition is the bog pimpernel *Anagallis tenella*, making a lacy pattern on the black peat with its spreading stems bordered with minute leaves, and pink funnel-shaped flowers raised on dainty stems. The rare marsh clubmoss *Lycopodium inundatum* is also found on bare areas of shallower acid peat in the slacks of Studland.

Late August and early September is a good time to see at its best what is perhaps one of the most beautiful plants of Dorset's damp acid heathland—the marsh gentian *Gentiana pneumonanthe*. Rising stiffly from the lush growth of the surrounding damp heath vegetation with narrow leaves shooting in opposite pairs up the stem, it holds aloft its azure blue trumpets so that the light shining through throws into sculptured relief the five distinctive green streaks on the outer side. In early autumn the main colour contrast comes from the orange fruiting heads of the bog asphodel so that the whole effect is particularly striking.

Early September, too, is the time when higher up on the drier heath west of Ferry Road one can see most clearly the overlap in Dorset of the western gorse *Ulex gallii* and the dwarf gorse *Ulex minor*. Both are in bloom growing side by side in a blaze of yellow, showing clearly the more compact growth of the dwarf gorse with its softer straighter spines and colder yellow flowers, with wings shorter than the keel. (In western gorse the wings are longer than or the same length as the keel.) The dwarf gorse fades out altogether further west, while the western gorse, except for its interesting appearance in the coastal belt of East Anglia, is mainly of western distribution.

DORSET HEATH ON HARTLAND MOOR

The clay that underlies and forms seams in the sands and gravels of the Hampshire Basin creates an impervious layer on which valley bog and wet heath is formed on the peaty waterlogged land. Extensive wet areas of this kind are found on the Hartland Moor and Morden Bog reserves. The humid heaths of the former are the main site in Britain for the Dorset heath *Erica ciliaris*. Research botanist S. B. Chapman of Furzebrook Research Station with whom I was fortunate to see Hartland Moor for the first time, has included among his pioneering studies of southern heathland, research into the distribution and habits of this plant—one of the loveliest of the heather family which is now confined to the Poole Basin of Dorset, a small scattered population in Cornwall and one untypical site on Dartmoor.

   Clustering at the top of the branching stem are its large globular flower-heads of a soft rose pink with style protruding from the narrow pouting mouth of the tubular corolla. The broadly oval leaves in whorls of three are fringed with hairs giving the plant its descriptive Latin name *Erica ciliaris* (fringed). According to S. B. Chapman (1975) this fringed leaved heath has what is known as a Lusitanian distribution in the world: north-west Morocco, coastal regions of Portugal, Spain, France, south-west England and western England. It is more upright in growth than the rather straggling cross-leaved heath *Erica tetralix* with which it hybridises so freely that a complete range of crosses have been found on the Dorset heaths and elsewhere.

DORSET HEATH
ERICA CILIARIS

175

On the northern edge of the Poole Basin heaths and south of the alluvial meadows of Seaford Water lies Morden Bog, once a large tract of heathland but now comprising 149 hectares (about 370 acres) of dry heath and bog surrounded on three sides by the Forestry Commission's Wareham Forest. This area has suffered less from uncontrolled fires than other heaths, and with its sharp transition from dry heath to bog, contains one of the few remaining extensive stands of fully grown heather *Calluna vulgaris*. All this makes it an important site for the scarce and quite harmless smooth snake that feeds mainly on lizards on heathland.

Writing on the smooth snake *Coronella austriaca*, Malcolm Smith (1951) quotes one of the older generation of naturalists who described how in 1868, when Bournemouth was still a small village surrounded by large expanses of moorland intersected by narrow valleys, the smooth snake was extraordinarily abundant. By 1888 the numbers had gradually decreased "and most of the wild moor having disappeared, they are not now met with in places where they formerly abounded. The favourite haunt is a dry sandy hillside overgrown with heather and gorse and coarse grass sloping down to a marshy valley where water is at all times available." This elegant snake with its sleek scales burrows freely in loose sand. It has been badly affected by plough-up and the fragmentation of heaths and is now almost entirely restricted to the heaths of Dorset, Hampshire and Surrey.

GRASSHOPPER LAND

Warm undisturbed heathland is a favourite haunt of grasshoppers, crickets and allied insects and for those interested in Orthoptera, the heaths of the Poole Basin are an excellent hunting ground, as D. R. Ragge (1965) points out in his invaluable book on the subject now unfortunately out of print. It was seeing a pink grasshopper matching the flowers of the heather on which it was resting, that first started an urge in Cyril Diver back in the early days of the Nature Conservancy to look at all the different variations in habitat on the South Haven Peninsula and find out what creatures lived there and how they adapted to the peculiarities of their environment. It was here that, as part of this search, he made a map of grasshopper distribution.

In the choice of a home, moisture and vegetation height often seem to be a critical factor. Hence the attraction of the richly varied Purbeck area with its high-low structure of vegetation.

The dry sandy Studland heathland suits the grey-green heath

grasshopper *Chorthippus vagans*, a very rare species found only on the heaths of Dorset and the extreme west of the New Forest. By contrast the greenish-yellow large marsh grasshopper *Stethophyma grossum* is an inhabitant of quaking bogs such as are found in the wettest parts of Morden Bog where it lays its egg pods at the base of tufts of purple moor-grass. This, the biggest of British grasshoppers, is also one of the rarer species.

Two groundhoppers among others found on the Dorset Heaths also have different moisture preferences. The common ground-hopper *Tetrix undulata*, well camouflaged on bare mossy soil, likes fairly dry ground with slightly damper places to hand so is admirably suited by the coastal heaths of Studland where dry ridges alternate with damper slacks. However, Cepero's groundhopper *Tetrix ceperoi* which also occurs on the Purbeck heaths, only makes a home near water into which it can jump and swim. It is at present restricted to the southern counties of England and Wales.

Moist areas where the cross-leaved heath and the purple moor-grass grow attract both the short-winged cone-head *Conocephalus dorsalis*, a small green bush-cricket with a yellowish or brown underside, and the bog bush-cricket *Metrioptera brachyptera* though both are sometimes found on drier ground.

A bog bush-cricket of the green variety was collected by permission on a Dorset wet heath late in its life cycle and brought to Suffolk for observation and drawing. It was a very spruce female with sickle shaped ovipositor, bright green margins to the forewings, green top to head and green "saddle" or pronotum with a broad pale band along the margin of the side flaps. Forewings were reduced to short lobes and hind wings almost non-existent.

It was observed in captivity from October 7th until its death on November 16th 1979 in an aquarium containing cross-leaved and Cornish heath growing in wet soil among seeding grass and sphagnum moss. It spent much time on the flower heads, leaping from one to the other but mainly crawling at low level among the vegetation, its green colouring exactly matching the narrow heather leaves. It rested sunning itself for long periods with its back to the sun on a bare patch. It was only seen to eat seed heads of grass and, though it may have eaten greenfly, did not take other plants offered and there was no sign of egg laying. In the wild, its food both meat and vegetable would have been more varied and it would probably have continued laying in a warm spell.

In comparison, a female dark bush-cricket *Pholidoptera griseo-aptera* adjusted much more readily to captivity, where it was observed over a longer time, from August 7th to November 21st 1979. It was dark brown in colour, larger and more strongly built with powerful hind legs and large curved ovipositor. Since this cricket was caught on a woodland fringe of heathland in Ashdown Forest among tall grass and blackberry bushes, its container was supplied with the same high vegetation, then placed alongside the bog bush-cricket in sunlight and watered regularly to avoid too much dry heat. Willow bark with deep crevices was provided for egg laying.

This cricket came immediately toward food offered, feeding voraciously first on dandelion leaf then almost totally on sow-thistle, biting through the calyx and eating the seeds and occasionally taking pear and apple. After two months with no meat protein, it ate straight off while watched a bluebottle, a mosquito and a medium sized moth (the streak), all of which were killed immediately before being dropped into the cage. After the first large feed, only an occasional greenfly and mosquito were accepted. Water drops were imbibed from sprinkled vegetation. Feeding took place at all times of day and late evening as did egg laying though in the wild this is not typical.

In the case of the dark bush-cricket, it was possible to watch egg laying in great detail. This was seen to take place at all times of the day from the beginning of September to mid November but it may well have taken place before this. Several attempts were made at first to drive the ovipositor into various small twigs but finally efforts were successfully concentrated on the willow bark. The body was raised into an arc with the ovipositor drawn down and forwards under the abdomen, swaying movements accompanying the exertion of pressing it downwards into the chosen crack. Prior to this in every case the bark surface was nibbled, small bits of bark being removed and thrown aside in what seemed to be a softening up process. After successful entry and egg laying, the bush-cricket slept for about half an hour before removing the ovipositor which towards the end of its life required quite a struggle. In one case, the cricket released itself with a jerk that sent it flying into the air and onto its back. Eggs are laid singly and are the shape of a white cigar pointed at both ends and some 2mm in size. The process was usually completed with a bite of food followed by a very thorough cleaning of the ovipositor which was drawn under the abdomen and forward to the mouth parts. All this took about $1\frac{1}{2}$ hours on at least three occasions. Eggs were still being laid four days before death on November 20th. On the last occasion three were laid on the surface.

Both the dark bush-cricket and the bog bush-cricket lost a hind leg the night of their death at the end of a typical life cycle. For some days previously they had been climbing laboriously around like very ancient creatures. Do they perhaps do this until a vital limb falls off and then die?

A female speckled bush-cricket *Leptophyes punctatissima*, apple green with purplish specks with a serrated upper edge to its ovipositor, climbed the studio window in Suffolk where the other two crickets were kept. It appeared to flourish in captivity where it sat on the taller blackberry leaves and gnawed the surface of the dried blackberry drupes. Since in the wild it lives in a similar habitat, it was put in the same large aquarium as the dark bush-cricket. One November night it disappeared. It may have been eaten by its companion as cannibalism does sometimes occur.

HEATHLANDS OF THE NEW FOREST

Though today the spatial links are few, the Dorset heaths are really a westward extension of the New Forest, lying in the same shallow Hampshire Basin. Similar aged sands and gravels give rise to heathland deep within the perambulation of the Forest.

"A poorer spot than this New Forest, there is not in all England; nor I believe in the whole world," wrote William Cobbett when making one of his rural rides in October 1826. In the eyes of Cobbett, the keen agriculturalist, barren heath was indeed a miserable waste for which no adjective was too strong. "Nothing can be more . . . villainously ugly . . . stony sand upon gravel . . . There is clay at the bottom of the gravel so you have nasty stagnant pools without fertility of soil."

Some scientists have argued that heath was the primeval condition of these infertile lands but Dimbleby (1962) and others believe that here as elsewhere it was man's clearance of relatively open woodland and the browsing of flocks that created the heathland. However, C. R. Tubbs (1969) points out that a high density of stock is needed to stop regeneration of trees altogether—about one bullock to four acres. This has led others to raise the question: had primitive people enough animals to keep down the shoots of seedling trees on the scale required? Whatever the answer, pollen records show that already in the new bronze age plains and woodlands were part of the Forest scene some thousands of years before William the Conqueror in the eleventh century named the New Forest and created the Forest Law to safeguard the royal venison.

179

My stay in the New Forest was based at Burley, a village chosen because of its nearness to Cranesmoor, a wonderful place for plants and insects attracted by its range of habitats from dry heath to valley bog.

Cranesmoor lies in a hollow where the surface geology is composed almost entirely of Barton Sands, and rain percolating over the centuries through this sandy soil has washed down mineral salts to form an impermeable iron pan at a depth of about 20–30cm. Water that cannot trickle through this layer ponds up on the surface, creating water-logged peaty conditions of valley bog. Prof. P. J. Newbould in 1960 made a detailed study of plant communities in this area and mapped their distribution in thirteen sites. One of his most interesting discoveries was the difference in plant composition where there was moving soil water compared with areas where there was no such flushing. This is because in the flushed area, even if the nutrient level is low it is constantly being replenished by moving water, whereas in the more stagnant sheltered parts this is not the case.

On Cranesmoor, the two lines of water flow are on the margins of the bog and here in the enriched less acid area one finds extensive communities of black bog-rush *Schoenus nigricans* with bog myrtle *Myrica gale*, meadow thistle *Cirsium dissectum*, and purple moor-grass *Molinia caerulea* where the soil water movement is greatest. At the other extreme in the central stagnant and more acid area there is a gradual change to sphagnum-rich vegetation with sundews *Drosera*, white beak-sedge *Rhynchospora alba* and in one or two places the rare tiny green-flowered bog-orchid *Hammarbya paludosa*.

THE PEAT MOSSES

Stagnant rainwater bogs contain few salts and other nutrients and waterlogging results in virtual exclusion of oxygen from the soil with the production of potentially toxic substances. However, sphagnum or peat moss is able to thrive in these conditions. It soaks up huge quantities of water by capillary action and holds it in the hollow cells of its leaf walls. As the sphagnum accumulates, it forms a layer of half decayed material which draws ground water upwards. From this dead sphagnum, peat is formed which when thoroughly wet is as impervious to water as dry rock.

Newbould's maps of individual bog moss varieties in Cranesmoor show that widespread in the stagnant area are the pale brownish-yellow hummocks of *Sphagnum papillosum*. Here, too, in the wettest part are the feathery yellowish green mats of *Sphagnum cuspitatum* while in dryer, shallower and less acid peat near the flushed areas the low gray-green cushions of *Sphagnum compactum* take over. One only has to squeeze these mosses like a sponge to see that they hold several times their weight in water. Because of its great absorbent power, sphagnum moss was collected and used for making absorbent dressings in both world wars.

FOREST GLADES AND PLAINS

After hours poring over the botanical treasures of a New Forest valley bog, it was a fine contrast to stretch one's legs for an hour or so among the great trees of the unenclosed forest, then at the height of its autumnal splendour. Now as in mediaeval times, one passes freely between wooded areas, forest glades and treeless plains—all part of the Crown waste comprised in the term *Forest*. This landscape pattern attracts many birds, particularly certain birds of prey that find here the mixture of open lowland heath and woodland they require. Some rare birds of prey find here in the New Forest, once one of the great game reserves, the haven they require for since the 1880s when the Hon. Gerald Lascelles took over as Deputy Surveyor of the Forest, there has been a deliberate policy towards bird of prey conservation and a tradition of tolerance towards them has been built up among the keeping staff which long predates the Protection of Birds Act 1954.

The hobby, a relatively scarce bird in Britain, clearly finds a refuge in the Forest. Colin R. Tubbs, Assistant Regional Officer (South) of the Nature Conservancy Council and author of an ecological history of the New Forest (1969), made a three year survey of its hobby population, and he tells me that today there are about sixteen pairs compared with the kestrel's twenty to thirty pairs. While feeding mainly on insects, this small falcon during the breeding season feeds its broods on small birds, often larks and pipits caught in spectacular stoops or in prolonged aerial chases over the open heaths and plains of the Forest.

The New Forest has its own somewhat isolated breeding population of buzzards which on fine days in spring and autumn can be seen soaring above the heathlands. These New Forest buzzards apparently include more bird flesh in their diet than is usual for their kind, though they also feed on small mammals of the woods and plains including grey squirrels, moles and young hares.

The honey buzzard, one of the two rarest breeding birds of prey, is not really a buzzard at all but it too requires large tracts of mixed woodlands with natural glades. Its main food consists not of honey but of wasp and bee grubs which it feeds to its young from the comb. Hampshire provides the honey buzzard with the summer warmth and the light sandy soil it requires where it can find wasp nests which it digs out with well adapted straight claws. The New Forest is the only place in Britain where a few have bred regularly.

Montagu's harriers, the smallest of our British harriers, disappeared from the New Forest in the early 1960s and from Britain as a whole in the mid seventies though a scatter of pairs have now reappeared in various parts of the country. Like the hen harrier, this is a bird of undisturbed heathland but at a lower, warmer level. The reasons for its decline are not clear but as Leslie Brown (1976) points out, if forestry eventually spread over all southern heathlands, the Montagu's harrier would disappear as a breeding bird from Britain. Mixed terrain with extensive heaths of the kind found in the New Forest is of the utmost importance for all these birds.

The great birds of prey share the freedom of the Forest with a number of large mammals notably badgers, four kinds of deer and of course the forest ponies. Prior to 1888, with the approach of the Winter Heyning or time of winter food shortage (November 22nd to May 4th) in theory all commoners' animals had to be withdrawn to ensure enough food for the royal deer, but now this restriction has been removed, and it was probably rarely observed.

In Burley Street and neighbourhood, all the year round the ponies have prior rights on the village streets as well as on the heaths and unenclosed forest. (Residents like my host find the need to change their gate latches from time to time as the ponies learn to open them.) Here in the Forest, commoners' rights over the Crown wastes are not just a nostalgic relic of bygone society, but of real economic importance. This applies particularly to two of these rights: the Common of pasture for commonable animals (which includes ponies, cattle, donkeys but not goats) and the Common of mast or pannage by which pigs may be turned out on the common to fatten on fallen acorns and beech nuts for a limited period of not less than sixty days. There is a separate Common of pasture for sheep but these rights are limited to properties once part of the estates of the Cistercian abbeys of Netley and Beaulieu, though a few sheep are in fact turned out under licence from the Verderers. The Common of estovers or free supply of wood fuel was still claimed by about a hundred or so houses in the Forest at the time of my visit but the right to dig marl and cut turf had more or less fallen into disuse. All these rights are attached to land not to persons and may be used by people occupying the land.

In the New Forest it is not only the three hundred year old trees that give a sense of great antiquity. This is our only forest where mediaeval institutions still survive as a framework for settling the affairs of the Forest community. Commoners and others with a problem to raise still take the matter before the historic Verderers' Court of Swainmote and Attachment that continues to meet as of old. A proportion of the Verderers are elected by the New Forest Commoners from their own number, a commoner for this purpose being the occupier of one acre of commonable land. Others are appointed by such bodies as the Forestry Commission, the local planning authority, Council for the Preservation of Rural England, and the Ministry of Agriculture.

Conflict between Crown and Commoners over the open wastes of the Forest stretches far back in time and during this struggle, the commoners have long formed their own defence associations. Today representatives of the New Forest Commoners' Defence Association attend the Verderers' Court and generally keep an eye on the welfare of the Forest.

There is a huge complex of commons within the perambulation of the Forest with some 44,505 acres, say 18,000 hectares of open forest over which common rights are exercised, only about 9,000 acres (3640 hectares) of which are woodland. In addition, there are about 3500 acres (1416 hectares) of manorial wastes under private freehold to which common rights also apply (Tubbs 1969). As the Forestry's publication *Explore the New Forest* shows, what was once the strictly preserved hunting ground of a Norman conqueror, as part of the process of history has become the shared inheritance of every citizen. With that heritage goes the solemn duty of keeping the Forest undamaged for future generations—no simple task when one realises that six million people are attracted there each year. It is however not so much the people as their technology that could pose perhaps a greater threat to this fabulous New Forest than at any other time in its nine hundred years' history.

THE HEATHS OF THE LIZARD PENINSULA

Except for the fact that both are strongly affected by their coastal position, perhaps the greatest contrast to the Sandling heaths of Suffolk in the east of England is found in the heathlands of the Lizard Peninsula of Cornwall in the extreme south west of this country. Large expanses of these heaths cover a plateau of land, ground flat by the elements, some three hundred feet at its highest point above the Atlantic into whose waters cliffs of spectacular beauty drop sheer on three sides. Salt-laden, gale-force winds of tremendous power blowing mainly from the west are present in varying degrees in every month and most weeks of the year. Over the centuries, these have resulted in the windpruned vegetation and comparative treelessness of the exposed west. Yet the Lizard like the rest of Cornwall has the warmest winters and one of the highest mean temperatures in the country with a freedom from severe frost that enables some plants to bloom all the year round. While in the widest sense most of Britain can be said to be oceanic in climate, here the climate is highly so, with that "Lusitanian" mixture of mildness and moisture that presents a considerable contrast to, for instance, the East Anglian climate.

CORNISH HEATH
ERICA VAGANS

Like so much in Cornwall, the heathlands have a character all their own. On Goonhilly in May of 1979 with the earth satellite station aerials afloat in high wind and glittering sunshine, one felt there could be no other place on earth quite like this flat windy land perched aloft under the immense arc of sky, but in fact some of these heaths of the Cornish serpentine rock have much in common with the heathlands off the coast of Brittany. Both share the plant communities of southern strongly oceanic regions where such plants as bell heather *Erica cinerea*, heath grass *Sieglingia decumbens*, tormentil *Potentilla erecta* grow together in an association often dominated by the beautiful Cornish heath *Erica vagans*. In Britain this hairless evergreen shrub with its whorls of 4–5 narrow rolled leaves is confined almost entirely to heaths near the Lizard where it is locally abundant. It thrives on brown earth of considerable depth and from July to September the long flowering spikes are packed with pink and lilac bloom from which chocolate-coloured anthers protrude conspicuously, making incomparable colour harmonies with peacock butterflies and other visiting insects. (White flowering Cornish heath is by no means rare.)

Most of our southern heaths are formed on acid soil that tends to produce a rather limited flora of acid-loving plants. What makes the Lizard heathland exciting to botanists is the high number of plants like the Cornish heath that flourish in neutral or alkaline conditions. Their occurrence here is due to the special nature of the underlying rocks which geologists describe as ancient igneous or metamorphic rocks of pre-Cambrian age. Below the heaths that spread over the central core of the western peninsula—like those on the Lizard and Goonhilly Downs—is a coarse crystalline rock of ferromagnesium silicate known as serpentine, an ornamental variety of which when highly polished to reveal its rich colouring and veining forms the basis of a local industry. In the field, serpentine rock disintegrates to a purplish-red clay, and the magnesium it contains causes water draining through it to become slightly alkaline. The main serpentine mass is surrounded by schists (sedimentary rocks) and gabbro (metamorphic rocks of aluminium silicates rich in calcium and magnesium) and intruded in the serpentine are areas of granite gneiss. Schist and gabbro give fertile, well-balanced soils nutritionally while serpentine is notoriously infertile, lacking calcium, potash and phosphate. So the contrasting rock types and their overlying soils, everywhere contaminated with wind-blown loess, give some interesting diversity to plant and communities.

By a most lucky chance, the visit to the Lizard heaths in May of 1979 coincided with a time when botanist Lewis C. Frost was in the area working with a colleague on some of its rare plants. His scientific paper *The Heaths of the Cornish Serpentine* (Frost and Coombe 1956) remains a classic work which for a lone explorer like myself was already proving a valuable aid in taking a closer look at the plant communities. A rare opportunity for questions and discussion kindly arranged by the Nature Conservancy Warden for the Lizard, Ray Lawman, could not therefore have been more timely.

The circumstances of our first chance meeting I well remember. On this particular May morning the turf of the rocky heath above Mullion Cliff was resplendent with the little blue spring squill *Scilla verna* that is not found on our East Anglian heaths, so walking at the cliff's edge, it was hard to choose between the dramatic sea-view of fulmars on the rocks below and the near view of these delightful small plants at one's feet.

The place was empty of people except for a group of three stretched out nose down in the short turf. This turned out to be the two botanists and the Lizard Reserves warden measuring a minute and extremely inconspicuous plant, *Isoetes histrix* or land quillwort which, with narrow leaves curved flat to the bare ground, looked rather like a little green terrestrial starfish. It grows only on the Lizard and the Channel Isles in Britain.

Several other very rare little plants mostly confined to the Lizard grow in the dwarf turf of these exposed heaths that cover the cliff tops. There is the unique genetical form of spring sandwort *Minuartia verna* adapted to the high magnesium content of the rock. In the dryest parts are three special Lizard clovers: twin-flowered, long-headed and upright *Trifolium bocconei, molinerii* and *strictum*. Another very rare plant not otherwise known in Europe except in the Channel Isles and possibly in France, is the fringed rupturewort *Herniaria ciliolata*. This neat little perennial survives the blasting of salt and wind it receives at the cliff's edge by growing in a mat close to the ground. Its leaves are fringed with minute silvery hairs which also cover the stem. The specimen illustrated was growing among wild thyme *Thymus drucei* where rock heath slopes down to the sea near Kynance. At the time, swards of pink thrift *Armeria maritima* clothed the whole slope and the sober pale green colouring of the rupturewort was nearly lost among the glowing orange-red and yellow of kidney vetch *Anthyllis vulneraria*, and the white of Danish scurvy grass *Cochlearia danica* and English stonecrop *Sedum anglicum*.

Maritime heath covers the cliff tops of the exposed coasts of the Lizard. In the most exposed places only a dwarf turf with wind-

FRINGED RUPTUREWORT *HERNIARIA CILIOLATA*

pruned bushes of heather *Calluna vulgaris* often with densely hairy leaves can survive but as shelter builds up inland, first bell heather *Erica cinerea* and then Cornish heath and gorse *Ulex europaeus* form a more dense and varied heathland. Rock heath is a variant of these maritime heaths and is one of the four natural associations of plants that Frost and Coombe recognised in their study among the many possible combinations of ten key plants. It develops on the exposed west coast at Mullion Cove, Gew Graze and Kynance, often extending to the cliff's edge on shallow well-drained reddish-brown loam with a pH value of about 6. The typical combination of heather *Calluna vulgaris* and some bell-heather with very little gorse of any type makes it easy to recognise. Of special interest to botanists are the genetically prostrate junipers and a prostrate form of dyer's green-weed *Genista tinctoria ssp. littoralis* growing here.

A feature of serpentine is that it can support basicoles, that is plants that require a high base status not necessarily calcium, as opposed to calcicoles which must have an adequate supply of calcium. Hence in the maritime heath at Kynance you will find the rare spotted cat's-ear *Hypochoeris maculata* and bloody geranium *Geranium sanguineum*, and in erosion pans on the cliffs the dark green clumps of wild chives *Allium schoenoprasum*, all of which are basicoles.

On rocky valley slopes on the serpentine like those at Mullion Cliffs, Gew Graze and Kynance, and on some of the more sheltered cliff slopes from Eastern Cliff to Black Head, a rather different type of heath occurs, very rich in species including an abundance of Cornish heath and gorse, both *Ulex europaeus* and *Ulex gallii*—a combination usually indicating deep brown earth and not found on the shallow rock heaths. In this type of heath, the vegetation varies so much from place to place that it was given the field name of "mixed heath".

Even on the Downs, where a serpentine outcrop protrudes tiny areas of "mixed heath" can be found. Flanking a boundary bank on the edge on the forest opposite the Goonhilly earth station is almost the sole remaining good example of mature mixed heath in which bushes of *Erica vagans* are aged from about fifteen to twenty years or more. This whole area is now a National Nature Reserve.

In the late summer and autumn, this older plant community is very striking when the flowering Cornish heath and the western gorse form a dense continuous mound of gold and lilac bloom but the adjoining stretches of burnt heath show how very nearly it was destroyed by the fierce fires of 1976. The destruction of a valuable and increasingly scarce habitat that takes many years to reach maturity and on which a whole complex of animal life depends is obviously a major tragedy and fire is a constant anxiety for any Reserve warden.

Anyone who has seen a Lizard fire in summer knows that the ferocity is so great that even the soil seems to be on fire. This is the surface humus being burnt up—something that not only impoverishes the soil but can lead to serious erosion by wind and water. Since most fires on the Lizard are tourist induced, all visitors have an immense collective responsibility to guard against the slightest carelessness in this connection.

Controlled fires at the lawful time of the year when the soil and vegetation are moist and therefore the burn is only superficial, do much less harm and indeed can add to the variety of species and quality for rough grazing. Cornish heath regenerates vigorously after a superficial burn and a temporary release of minerals gives rise to a luxuriance of herbs on the burnt heaths with carpets of oxeye daisies *Leucanthemum vulgare* and bog pimpernel *Anagallis tenella.*

TALL HEATH AND SHORT HEATH

The casual visitor standing in the middle of Goonhilly Downs cannot but notice that the flat plateau of the landscape is varied with slight mounds of darker vegetation surrounded by shallow hollows of lighter vegetation that act as drainage areas. These mounds are composed of very acid loess (pH 4 at the surface)—a silty material of granitic origin which was wind transported from the granite masses of West Cornwall probably at the end of the last glaciation. The acid loess gives rise to "short heath", while the drainage hollows receiving water from the serpentine are neutral to alkaline and support "tall heath".

In the last fifteen years, considerable areas of Lizard heathland have been ploughed up for the first time in their history. It was therefore an important step when in 1976–77, two hundred and six acres of wet heath on Goonhilly that has probably never been enclosed or improved became Cornwall's first National Nature Reserve, purchased for this purpose by the Nature Conservancy Council. Where the land dips into a hollow and pans of peaty water between clumps of purple moor-grass *Molinia caerulea* make walking difficult, one notices immediately the special feature of this wet area—the tall pale stems and black inflorescence of the bog rush

Schoenus nigricans giving an impression of height to the vegetation. The dominant plants of this so-called tall heath association are black bog rush and Cornish heath. By perching clear of the general moisture on a small outcrop of serpentine rock it was possible also to see the small broom-like yellow flowers of petty whin *Genista anglica* with its spiny, dark green wiry stems, the saw-toothed leaves of saw-wort *Serratula tinctoria* that likes alkaline soils but will grow on more acid ones, the stiff bluish-grey leaves of carnation sedge *Carex panicea* not found on very acid soils, the somewhat straggling growth of cross-leaved heath *Erica tetralix* and the buttercup yellow four-petalled flowers of tormentil *Potentilla erecta*, both plants of acid heaths but with a wide tolerance of soils. The greater burnet *Sanguisorba officinalis*, which we see very rarely in Suffolk, was obviously flourishing in the moist heath but in May had not yet developed the oblong heads of small dark red flowers. These deep waterlogged soils of the serpentine where the bog rush combination of tall heath develops are more alkaline the further one penetrates to the serpentine rock, varying from pH 5–6 near the surface to over 7 nearer the rock. Hence perhaps the combination of both acid-loving and base-loving plants.

In the short heath of the raised acid areas, heather, bell heather and western gorse replace the bog-rush and Cornish heath of the lower, wetter ground. The prevalent grass is the grey-green bristle bent *Agrostis setacea* (locally very abundant in these south-western heathlands) with a certain amount of purple moor-grass *Molinia caerulea* but no sheep's fescue *Festuca ovina*. In the early years of this century the surface of these heaths was pared for turf. Today signs of turf cutting and the ruts of carts are still visible in uneven surfaces where water stands in winter, drying out in summer to form a habitat for rare little rushes like the pygmy rush *Juncus pygmaeus*.

VIOLETS AND ORCHIDS

Violets and milkworts are plants of this short heath community and one of the most pleasant memories of mid May 1979 was the sight of sunlight after rain falling on a profusion of violets growing in the fine grass among the dark blue and dark pink of the heath milkwort *Polygala serpyllifolia*. Crosses between different species of violets produced a bewildering variety of colour and form making generalisation difficult. However, with help it was possible to distinguish with reasonable certainty the pale milky blue of the pale violet *Viola lactea*, mainly restricted to south-western heaths, with its short green spur and narrow, often wedge-shaped leaves; and what the books call the violet blue of the common dog violet *Viola riviniana*, with its curved whitish spur, broad heart-shaped leaves and characteristic central flowering rosette. (The heath violet *Viola canina* does not grow on the Lizard.)

The dog violets attract both the small pearl-bordered fritillary butterfly and the dark green fritillary to the area. Both butterflies lay their eggs on the leaves of these plants. The fully grown hairy caterpillars of the first, *Boloria selene*, are described by South as smoky-pink and velvety looking, while those of the dark green fritillary *Argynnis aglaja* are "a purplish-grey mixed with velvety black", with a yellow stripe down the middle.

Orchids are another delight of the Lizard heaths and are best seen perhaps on the Crousa Downs, a haunt of stonechats where boulder-strewn heathland covers the gravel and the underlying greenish grey Gabbro rock. One of the first to flower is the green-veined orchid *Orchis morio*, so called from the parallel green veins on the purple sepals. Later, from June to August, the heath spotted-orchid *Dactylorhiza maculata*, an orchid of damp, fairly acid heaths, will be raising broad spikes of pale pink and purple flowers from its spotted leaves. One must look in the areas of short heath for the sweet-smelling flowers and long slender spurs of the fragrant orchid *Gymnadenia conopsea*—an orchid that comes into bloom from June to July and prefers fairly alkaline soils.

FAUNA OF THE LIZARD HEATHLANDS

Most of the common breeding birds of south-west heathlands have been recorded as nesting on the Lizard in the following approximate order of abundance:—skylark, meadow pipit, stonechat, yellow-hammer and whitethroat. Wheatears are scarce coastal nesters, while one, sometimes two, pairs of nightjars still nest annually in one unspecified place on the Lizard. Curlews and lapwings also breed in small numbers on the heaths.

For this information I am indebted to correspondence from R. D. Penhallurick, assistant curator of the County Museum, Truro and author of *The Birds of Cornwall and the Isles of Scilly* (1978) etc, who tells me that though it is logical to assume breeding in the eighteenth century, there is no reference to Dartford warblers ever having nested nearer the Lizard than Constantine on the Carnmenellis granite

190

where they were plentiful in the 1970s. Even as passage birds on the Lizard, there have been only two definite sightings both at Kennack on April 14th 1940 and February 3rd 1974, though others may well have been missed.

Of birds of prey, kestrels and buzzards are both commonly seen and known to nest on the Lizard. The Montagu's harrier probably bred regularly until the nineteenth century but has much decreased in the last hundred years, the only records from the Lizard peninsula being for 1911 or 1912, perhaps 1931, probably 1939 and certainly 1968. Otherwise this harrier is now only a scarce passage migrant, on the Lizard, though in other parts of Cornwall it has bred probably annually up to and including the 1960s and occasionally since, in recent years preferring young Forestry Commission plantations to the more "natural sites" of deep heather and western gorse.

Based on the observations of Robin Khan in mid-Cornwall reported to R. D. Penhallurick in the early 1960s, the Montagu's harrier's main food here in order of preference appears to be short-tailed voles, common lizard, beetle spp., slowworm, common frog and the juveniles of grasshopper warbler, willow warbler, chiffchaff and stonechat.

MERLIN

Of another moorland breeding bird of prey, the merlin, there are no nesting records in Cornwall at any period except one solitary occurrence in 1954. In winter this smallest British falcon comes annually to the Cornish moors but must be described as scarce, a stature given it by Couch in 1938.

Of the hen harrier R. D. Penhallurick writes: "I have my doubts that this species ever nested in Cornwall, let alone the Lizard. The 'furze kits' of the countrymen to my mind referred to Montagu's harriers when they referred to harriers at all." Nor was there in 1978 any evidence of the regular wintering of the hen harrier on the Lizard, in contrast to the Land's End peninsula where it takes up territories as a regular wintering bird in small numbers. For the Lizard, the few published sightings point to spring and autumn migrants only.

To date, 2000 acres (810 hectares) of scientifically important areas of the Lizard have been established as Nature Reserves by the Nature Conservancy Council, the National Trust and the Cornwall Naturalist Trust.

A Handbook of the Natural History of the Lizard Peninsula (F. A. and S. M. Turk 1976) besides giving a useful list of flowering plants, contains information concerning the fauna of the various habitats including the Reserves. Of this it is only possible to indicate a few points of special interest.

While there are very few rabbits on the heaths, the brown hare *Lepus capensis occidentalis* finds the flat lands of the Lizard peninsula to its liking and is common there. Numbers have been increased in recent years by the introduction of both the brown hare and the shorter eared blue hare *Lepus timidus scoticus*. Badgers on the Lizard in the absence of trees make their homes in broken areas of cliff on the rock heaths. Reptiles include a melanistic form of adder known locally as "the croft adder". It is found in dry ditches in the depths of the heathland and the usual zigzag markings are scarcely distinguishable.

The richness of insect life on the peninsula may be hinted at by one of the better recorded groups, the shieldbugs and their allies, most of which are plant juice feeders. Of the 129 species so far recorded *Myrmedobia bedwelli*, a species described as new to science, was collected on wild thyme *Thymus drucei* at Kynance in 1933 but never found since. Bristowe (1935) recalls a similar instance of a very rare spider being found in the same area but not seen there since. It was almost certainly *Eresus niger*, the male having a bright scarlet abdomen—a spider that we know is not extinct since it has been found recently in Dorset.

These and other fascinating matters can be further pursued in the excellent publications of the Lizard Field Club, including their annual publication *Lizard*.

Despite all the immense pressures of the present day according to the County Council Structure Plan (Cornwall's seasonal influx was three and a half million in 1976), here on the Lizard peninsula intractable soil and exposed position still give rise to some of the least fragmented or otherwise altered areas of lowland heath in southern Britain and on that hopeful but unfortunately untypical note must end this account of one person's search for a habitat that nationally speaking can be seen to be accurately described as *our vanishing heathlands*.

READING REFERENCES

Alford, D. V. 1975 *Bumblebees* Davis-Poynter

Alvin, D. L. 1977 *The Observer's Book of Lichens* Warne

Armstrong, P. H. July 1971 "The Heathlands of the East Suffolk Sandlings" *Suffolk National History* Vol. 15 Pt 5

Axell, H. and Hosking, E. 1977 *Minsmere: Portrait of a Bird Reserve* Hutchinson

Bannister, P. 1965 "Biological Flora of British Isles. *Erica cinerea" Journal of Ecology* 53, 527–542

Berry, R. 1979 "Nightjar Habitats and Breeding in East Anglia" *British Birds* Vol. 72 Nos: 207–218

Bewick, T. 1862 *A Memoir of Thomas Bewick, written by Himself* ed. Jane Bewick See also abridged text ed. M. Weekley 1961 The Cresset Press

Bibby, C. 1976 "Bird on a Knife-edge" *Birds* (R.S.P.B. magazine) Vol. 6 No. 5 43–4

Blackstone, W. 1765 *Commentaries on the Laws of England* Clarendon Press

Bloomfield, R. 1800 *The Farmer's Boy* Capel Lofft

Boreham, H. J. 1951 "Life History of the Digger Wasp, *Mellinus arvensis, L.*" *Transactions of the Suffolk Naturalists' Society* Vol. VII–Part III p. 98–107

Bradley, J. D. and Fletcher, D. S. 1979 *British Butterflies and Moths* (with updated nomenclature) Curwen

Brian, M., Abbott, A., Pearson, B. and Wardlaw, J. 1977 *Ant Research 1964–76* Institue for Terrestrial Ecology, Furzebrook Research Station

Brian, M. V. 1977 *Ants* Collins N.N.

Briggs, A. 1965 *Chartist Studies* Macmillan

Brightman, F. H. and Nicholson, B. E. 1979 *The Oxford Book of Flowerless Plants* Oxford University Press

Bristowe, W. S. 1976 *World of Spiders* Collins N.N.

Brown, L. 1976 *British Birds of Prey* Collins N.N.

Burrell, E. D. R. 1960 "An Historical Geography of the Sandlings of Suffolk 1600–1850" London University M.Sc. thesis Suffolk Record Office (qv)

Burton, M. 1976 *Guide to the Mammals of Britain and Europe* Elsevier Phaidon, Oxford.

Chapman, S. B. and Webb, N. R. 1978 "The Productivity of a Calluna Heathland in Southern England" *Ecological Studies*, Vol. 27 Springer-Verlag, Berlin, pp. 247–62

Chapman, S. B. 1975 "The distribution and composition of hybrid populations of *Erica ciliaris* and *Erica tetralix* in Dorset" *Journal of Ecology* 63, 809–824

Chapman, V. J. 1964 *Coastal Vegetation* Pergamon Press

Chatwin, C. P. and Norton, P. E. P. 1974 "The Icenian Crag of South-east Suffolk" *Philosophical Transactions of the Royal Society of London* B. 269, 1–28

Chinery, M. 1973 *A Field Guide to the Insects of Britain and Northern Europe* Collins

Clapham, A. R., Tutin, T. G. and Warburg, E. F. 1962 *Flora of the British Isles* 2nd Edn. Cambridge University Press

Clare, J. 1821 *Poems Descriptive of Rural Life and Scenes, by a Northampton-shire Peasant*

Claridge, J. 1670 *The Shepherd of Banbury's Rules* Reprint Sylvan Press 1946

Clarke, W. G. 1925 *In Breckland Wilds* London: Robert Scott

Cobbett, W. 1821 *Rural Rides* Everyman, Vol. 1

Crabbe, G. 1810 *The Borough* John Hatchard

Culpeper, N. 1653 *Culpeper's Complete Herbal* Enlarged edn. Richard Evans London 1815

Daniel, W. B. 1812 *Rural Sports* Longman, London

Darby, H. C. 1971 *The Domesday Geography of Eastern England* Cambridge University Press

Darlington, A. 1978 *Mountains and Moorlands* George Rainbird Ltd

Defoe, D. 1724 *A Tour through England and Wales* reprinted in Everyman's Library 1928. Vol. 1

Delaney, M. J. 1956 "The Animal Communities of three areas of pioneer Heath in S.W. England" *Journal of Animal Ecology* 25, 112–26

De la Rochefoucauld, F. 1784 *A Frenchman in England* ed. Roberts, S. C. 1933 Cambridge University Press

Dempster, J. P. and Lakhani, K. H. 1979 "A Population Model for Cinnabar Moth and its food plant, ragwort" *Journal of Animal Ecology* 48, 143–163

Denman, D. R., Roberts, R. A. and Smith, J. H. F. 1967 *Commons and Village Greens* Leonard Hill

Dimbleby, G. W. 1962 "The development of British Heathlands and their Soils" *Oxford For. Mem.*, 23, 1–121

Diver, C. 1892–1969 *A Memoir* Furzebrook Research Station 1971

Dony, J. G., Rob, C. M. and Perring, F. H. *English Names of Wild Flowers* Botanical Society of the British Isles Butterworth

Else, G., Felton, J. and Stubbs, A. 1978 *The Conservation of Bees and Wasps* Nature Conservancy Council

Evans, G. Ewart. 1956 *Ask the Fellows who Cut the Hay* Faber

Fabre, J. H. *Trans. Teixeira de Mattos.* 1915 *Bramble bees and Others* Hodder and Stoughton

—— 1915 *The Sacred Beetle and Others* Hodder and Stoughton

—— 1919 *Slowworm and other Beetles* Hodder and Stoughton

Farrow, E. P. 1925 *Plant life on East Anglian Heaths* Cambridge University Press

Ferns, P. N. 1979 "Growth, Reproduction and Residency in a Declining Population of *Microtus agrestis*" *Journal of Animal Ecology* 48 739–75

Forestry Commission 1975 *Explore the New Forest* H.M.S.O.

Free, J. B. 1959 *Bumblebees* Collins N.N.

Frost, L. C. and Coombe, D. E. 1956 "The Heaths of the Cornish Serpentine" *Journal of Ecology* 44 226–256

—— 1956 "The Nature and Origin of the Soils over the Cornish Serpentine" *Journal of Ecology* 44 605–615

Frost, L. C. 1968 "Rare Plants of the Lizard District" *Lizard 3 (4).* (Published by Lizard Field Club)

Furzebrook Research Station, *21 Years of Research* Institute of Terrestrial Ecology, N.E.R.C.

Gimingham, C. H. and Barclay-Estrup. 1969–71 "The Description and Interpretation of Cyclical Processes in a Heath Community" *Journal of Ecology* 57 737–58; 58 243–9; 59 143–66

Gimingham, C. H. 1972 *Ecology of Heathlands* Chapman and Hall

Gimingham, C. H. 1975 *An Introduction to Heathland Ecology* Oliver and Boyd

Gimingham, C. H. 1960 *Calluna Vulgaris (L)*. Biological Flora of the British Isles *Journal of Ecology*

Gimingham, C. H. 1949 "Effects of grazing on balance between *Erica Cinerea* (L) and *Calluna Vulgaris* (L) Hull in Upland Heath." *Journal of Ecology* 37 100–119

Glyde, J. 1866 *The New Suffolk Garland*

Godfrey, G. K. 1953 "The food of *Microtus agrestis*." *Sangetierk Mitt* 1 148–151

Godwin, H. 1975 *The History of the British Flora second edition* Cambridge University Press

Goodwin, W. 1785 *Diaries* Suffolk Record Office (qv)

Hagerup, O. 1950 *Thrips Pollination in* Calluna. 1 Kommission HOS EJ NAR Munksgaard, Copenhagen.

Hammond, C. O. 1979 *The Dragonflies of Great Britain and Ireland* Curwen

Hardy, T. 1896 *The Woodlanders* Osgood, McIvaine and Co.

Hardy T. 1895 *The Return of the Native* Osgood, McIvaine and Co.

Haslam, I. 1954 "Myxomatosis in East Suffolk". *Transactions of the Suffolk Naturalists Society*. Vol. IX Part III

Hepburn, I. 1952 *Flowers of the Coast* Collins, *New Naturalist*

Hey, R. G. 1967 "The Westleton Beds Reconsidered." *Geological Proceedings* Vol. 78 p. 427

Hilton, R. 1977 *Bond Men Made Free* Methuen

Hind, W. M. 1889 *Flora of Suffolk* Gurney and Jackson

Hoskins, W. G. and Stamp, Dudley 1963 *The Common Lands of England and Wales* Collins

Imms, A. D. 1947 *Insect Natural History*. Collins, *New Naturalist*

Kirby, J. 1753 *The Suffolk Traveller*. (second edition 1764)

Lack, D. 1935 "Breeding Bird Population of British Heaths and Moorland." *Journal of Animal Ecology* 4 43–51

Lack, D. 1933 "Habitat Selection in birds." *Journal of Animal Ecology* 2

Lack, D. 1932 "Some breeding habits of the European Nightjar." *IBIS 74:* 266–238

Lennard, R. 1945 "The Destruction of Woodland in the Eastern Counties under William the Conqueror." *Economic History Review* XV

Linssen, E. E. 1959 *Beetles of the British Isles* Vol 1 and 2 Warne

Lockley, R. M. 1964 *The Private Life of the Rabbit* Andre Deutsch (Corgi reprint 1973)

Lomholdt, O. 1975–76 "The Sphecidae (Hymenoptera)" Vols I and II *Fauna Entomologica Scandinavica* (In English)

Long, D. C. 1959 "Some Non-Marine Mollusca from the Icenian." *Transactions Suffolk Naturalists' Society* Vol. XI Part II

Longfield, C. 1949 *Dragonflies of the British Isles* Warne

Main, H. 1917 "Pupation of Dysticus marginalis and Geotropus typhoeus." *Proceedings of the Entomological Society, London*, 1917

Matthews, L. Harrison 1952 *British Mammals*, Collins *New Naturalist*

McClintock and Fitter 1967 *Pocket Guide to Wild Flowers* Collins

Mendel, H. 1978 "Nebrina salina (Fairemaire) in Suffolk." *Suffolk Natural History* 17 407–8

Merrett, P. 1967 "The Phenology of Spiders on Heathland in Dorset." *Journal of Animal Ecology* 36, 363–374

Moore, N. W. 1962 "The Heaths of Dorset and their Conservation." *Journal of Ecology* 50, 369–391

Morton, A. L. 1945 *A People's History* Lawrence and Wishart

Nature in Norfolk: A Heritage in Trust Jarrold

Newbould, P. J. 1960 "The Ecology of Cranesmoor, a New Valley Bog." *Journal of Ecology* 48 361–383

Ogilvie, F. M. 1920 *Field Observations on British Birds* Selwyn and Blount

Payn, W. H. 1978 *The Birds of Suffolk* Ancient House Publishing, Ipswich.

Peacock, A. J. 1965 *Bread or Blood* Gollancz

Pearson, D. J. 1973 "Changes in the bird population on Walberswick Heaths." *Suffolk Natural History* Vol. 16 Part 3, p. 152

Penhallurick, R. D. 1969 *Birds of the Cornish Coast* Penzance Headland Press

Penhallurick, R. D. 1978 *The Birds of Cornwall and the Isles of Scilly* Penzance Headland Press

Pennington, W. 1974 *A History of Vegetation* The English University Press

Postgate, M. S. 1962 "The Field Systems of Breckland." *Agriculture History Review* Vol X Part II 80–101

Presst, I. 1971 "An Ecological Study of the Viper in Southern England." *Journal of Ecology* 16 pp. 373–418

Rackham, O. 1976 *Trees and Woodland in the British Landscape* Dent

Ragge, D. R. 1965 *Grasshoppers, Crickets and Cockroaches of the British Isles* Warne

Rainbird, W. and H. 1849 *On the Agriculture of Suffolk* Longman

Reyce, R. 1618 *The Breviary of Suffolk*. Ed. Hervey Lord, F. 1903 (Suffolk in the Seventeenth Century.)

Salisbury, E. J. 1931–2 "The East Anglian Flora." *Transactions of the Norfolk and Norwich Naturalists' Society* 13

Simpson, F. W. 1965 "Flora of the Suffolk Crag." *Transactions of the Suffolk Naturalists' Society* vol 13 part I pp. 1–10

Smith, A. c. 1860 "In City Streets." *The Open Road Anthology* Ed E. V. Lucas 1899 Grant Richards

Smith, M. *The British Amphibians and Reptiles* Collins *New Naturalist*

South, R. 1977 *Moths of the British Isles* Edited and revised by Edilsten, Fletcher and Collins Vol 1 and 2 Warne

South's *British Butterflies* Ed. T. G. Howarth 1973 Warne

Spencer, H. E. P. 1979 "A Contribution to the Geological History of Suffolk." *Suffolk Naturalists' Society*

Step, E. 1946 *Bees, Wasps, Ants and Allied Insects* Warne

Stephen, H. J. 1841–45 *New Commentaries on the Laws of England* Butterworth

Street, D. 1979 *Reptiles of Northern and Central Europe* Batsford

196

Suffolk Domesday Translated 1889

Suckling, A. 1846 and 1848 *Antiquities and History of Suffolk* John Weale

Tansley, A. G. 1911 *Types of British Vegetation* Cambridge United Press

Tansley, A. G. 1968 *Britain's Green Mantle* Revised and edited by Proctor C. F. Allen and Unwin

Tate, P. 1977 *East Anglia and its Birds* Witherby

Tate, W. E. 1951 "Handlist of Suffolk Enclosures and Awards" *Proceedings of the Suffolk Institute of Archaeology* Vol. 25. part 3. pp. 225–63

Thirsk, J. and Imray, J. 1958 "Suffolk farming in the 19th Century." *Suffolk Records Society* Vol 1

Ticehurst, C. B. 1932 *A History of the Birds of Suffolk* Gurney & Jackson

Trist, P. J. O. 1979 *An Ecological Flora of Breckland* E.P. Publishing Ltd

Tubbs, C. R. 1969 *The New Forest: an Ecological History* David and Charles

Turk, F. A. and Turk, S. M. 1976 *A Handbook to the Natural History of the Lizard Peninsula* Department of Extra Mural Studies, University of Exeter

Turner, C. 1970 "The Middle Pleistocene deposits at Marks Tey, Essex." *Philosophical Transactions of the Royal Society of London* B. 257, 373–440

Tusser, T. 1557 *A Hundreth Good pointes of husbandrie. 1573 Five hundreth Points etc.* In British Library. See also, Thomas Tusser. Ed. D. Hartley (1931) *Country Life*

Wainwright, G. J. 1972 "An excavation of a neolithic settlement in Brome Heath, Ditchingham, Norfolk." *Proceedings of the Prehistoric Society* Vol III

Watling, R. 1973 *Identification of the Larger Fungi* Hulton

Watt, A. S. 1955 "Bracken versus heather, a study in plant sociology," *Journal of Ecology* 49 709–15

Webb, N. R. 1980 "The Dorset Heathlands: Present Status and Conservation" *Bulletin of Ecology* t. 11, 3 pp. 659–664

West, R. G. 1955 "The Quaternary Deposits at Hoxne, Suffolk and their Archaeology." *Proceedings of the Prehistoric Society* 20 131–154

West, R. G. "East Anglia" *INQUA 1977* Geo. Abstracts Ltd. University of East Anglia, Norwich

West, R. G. and Norton, P. E. P. 1974 "The Icenian Crag of Southeast Suffolk." *Philosophical Transactions of the Royal Society of London* B 269

White, W. 1844 and 1874 *History, Gazetteer and Directory of Suffolk* William White, Sheffield

Young, A. 1771 *The Farmer's tour through the East of England.* London

Young, A. 1794 *General View of the Agriculture of the County of Suffolk. Second edition 1797 Third Edition 1797, 1804 Reprint* David and Charles 1969

Blois Family Archives: S.R.O. HA 30: 50/22/3.1 (Commoners of Walberswick petition 1637; Lease of Westwood Lodge 1770)

Burrell, E. D. R. 1960. "An Historical Geography of the Sandlings 1600–1850." S.R.O. 722

Goodwin, W. 1785 *Diaries* S.R.O. H.D. 365

Leiston Tithe Award and Map 1841: S.R.O.: FDA 164/A1/1a, 1b

Leiston Common Enclosure Award 1824: S.R.O. EF 5/1/2/1

INDEX

Entries for individual species are grouped under general headings e.g. plants; moths; mammals. Where appropriate, each is listed under its common name and its Latin name.

210

215

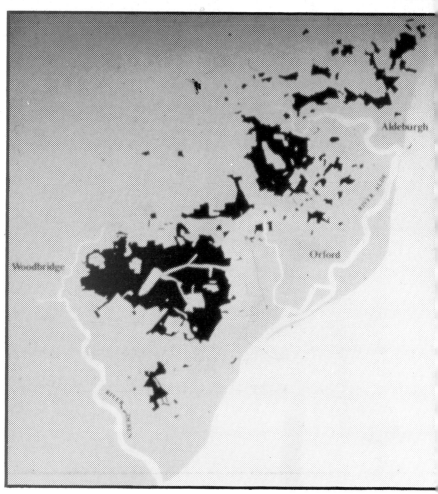

1881

THE LOSS OF HEATHLAND

SANDLINGS

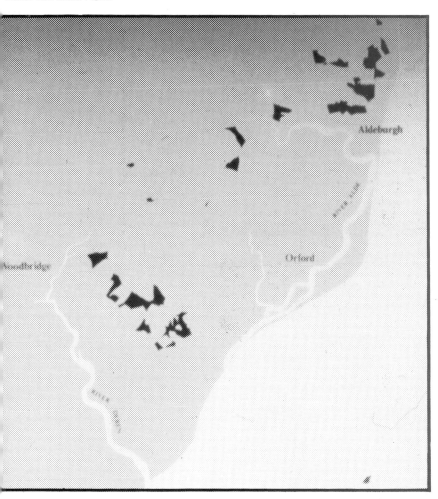

1981

OVER THE LAST HUNDRED YEARS